The Spad A Series

Jack Herris & David Méchin

Great War Aviation Centennial Series #104

Publisher's Note

For our aviation books, please see our website at: **www.aeronautbooks.com**.
I am looking for photographs of the less well-known German aircraft of WWI to complete this series. For questions or to help with photographs you may contact me at **jherris@me.com**

ISBN: 978-1-964637-49-5

Color Profiles: David Méchin
Design and layout: Jack Herris
Cover photo colorized: Alexander Feosenko
Cover design: Aaron Weaver
Digital photo editing: Jack Herris

Aeronaut Books
Books for Enthusiasts by Enthusiasts
www.aeronautbooks.com

Table of Contents

Acknowledgements

Much of this material was originally published in French aviation magazines by Christophe Cony. Martin Digmayer suppled the scale drawings. Photos were provided by James Davilla, Jon Guttman, Colin Owers, Walter Pieters, and the late Ellic Sommer among others.

Notes on SPAD S.A.Series Aircraft

Type	SFA Serials	Engine	Remarks (Armament, etc.)
Prototype	Renumbered to SFA S-11	80 hp Le Rhône 9C	Originally unarmed, then modified into S.A.1 configuration with flexible gun. Tailplane was trapezoidal. Ailerons on all wings,
S.A.1	S-! to S-11	80 hp Le Rhône 9C	Early without additional cooling holes in forward fuselage. At least S-4 and S-7 had the enlarged cooling ducts of the S.A.2. S-4 was retrofitted with dual-tube gun mount. Ailerons on all wings,
S.A.2	S-12 to S-39 S-41 to S-47	110 hp Le Rhône 9J	More powerful engine had larger cooling vents, dual-tube gun mount. Ailerons on all wings,
S.A.3	S-40	110 hp Le Rhône 9J	Flexible gun added for pilot, dual controls, forward nacelle widened for rudder bar contols, Ailerons on all wings,
S.A.4	S-48 to S-96	80 hp Le Rhône 9C 110 hp Le Rhône 9J	Airframe like S.A.2 but ailerons on upper wing only and a cabane center section. Most had the 80 hp engine but ten with the 110 hp engine.

Introduction

The story begins with Armand Deperdussin. A wealthy Parisian silk broker, he founded an airplane manufacturing company in 1909 and hired a young 29-year-old engineer, Louis Béchereau, a former associate of the renowned Clément Ader. The company distinguished itself by producing monoplane racing aircraft with monocoque fuselages, which, as early as January 1912, broke the world speed record, exceeding 145 km/h, and then surpassing 200 km/h in September 1913. However, while Deperdussin's airplanes enjoyed continued success, the company itself was faltering. Armand Deperdussin had mismanaged his assets and had embezzled money from one of his companies. He was declared bankrupt in early 1913 and imprisoned, his property being sold at auction.

Louis Blériot, the aviation pioneer who made his fortune by successfully crossing the English Channel in 1908 with his Blériot XI, was by then no longer young (42 years old) and somewhat behind the times in terms of technology. He understood that for the long-term viability of his company, he needed to hire new talent. With the bankruptcy of the Deperdussin company, he sensed a good opportunity thanks to the expertise of the engineer Béchereau, and bought the company on the eve of the First World War, together with a few industrialist friends.

Above: The AEG G.II was designed as an aerial cruiser with the ability to carry bombs. Operational experience soon showed it was far more effective as a bomber than in aerial combat as an interceptor and future AEG twin-engine aircraft were designed and used as bombers.

The company was managed by the court-appointed liquidator, a certain Raynaud, who formed the Société Provisoire Armand Deperdussin, SPAD. At the first board meeting, the question arose of what to name the new company. Alfred Leblanc, a friend of Blériot, suggested keeping the name SPAD. Why? In Esperanto, a language

Above: The LVG G.I was designed as a *Kampfflugzeug*, and a single example was built in 1915. The arrangement enabled the gunner to fire in a 360° arc above the aircraft and gave a wide field of fire forward, but the standing gunner undoubtedly created a great deal of drag, limiting the aircraft's speed. The *Kampfflugzeugs* (battle planes) of all types soon proved failures as interceptors and air superiority aircraft because they were simply too slow and cumbersome to intercept and engage enemy aircraft. Operational experience soon demonstrated that bombing was the true role of these aircraft, and the early operational *Kampfflugzeugs* were the ancestors of German twin-engine bombers of WWI.

This Page: A 1913 design for an aerial cruiser, the French Dorand Armored Interceptor had two 80hp engines coupled together to power both propellers in case of an engine failure. There was a pilot in the middle with gunners fore and aft. Development of armored aircraft was abandoned because of poor performance due to the armor's weight.

Left: The Vickers F.B.5 Gunbus was designed to attack enemy aircraft with its flexible gun. Despite its relatively low speed it had some success due to the field of view and field of fire enabled by its pusher configuration.

Above: The Germania DB was similar in concept to the AEG C.III. Designed as a two-seat fighter, the gunner sat in the front cockpit and was armed with a flexible and unsynchronized machine gun, with the propeller obscuring the field of fire forward and downward. There was no synchronized gun for the pilot. The DB also featured a fish-shaped fuselage profile with rectangular cross section for streamlining, but had external radiators. It was a 2-bay design that remained a prototype.

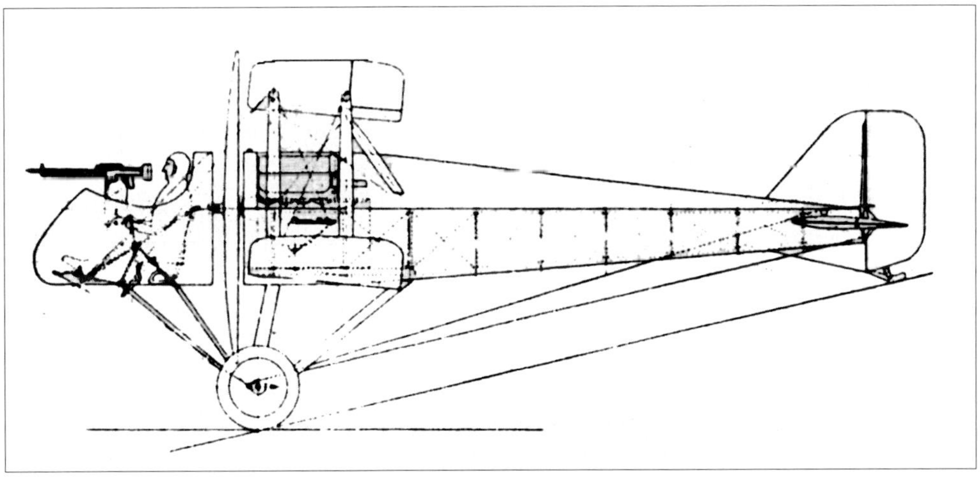

Right: The S.E.10 design was similar to the successful S.E.5 but was never built.

that was currently fashionable and very popular in the aeronautical world, SPAD means "speed," being derived from the English word. So, SPAD it is, it sounds great! Officially, they've found a new acronym: Société Pour l'Aviation et ses Dérivés (Society for Aviation and its Derivatives).

Right: The Gotha G.I was designed as a multi-purpose battleplane that could carry bomb. Too slow to successfully intercept enemy aircraft, it was used for bombing and sired the famous Gotha bomber series.

This Page: Developed from the B.E.2, the B.E.9 featured the same 'pulpit' gunner's cockpit configuration as the Spad S.A.series. B.E.9 serial 1700 was sent for operational testing in France in September 1915; Capt Robin Rowell test flew this and said he was happy it was crashed. (lower Greg VanWyngarden)

Above: The Deperdussin TT was another two-seat fighter design with pilot and gunner with flexible, unsynchronized gun. The gunner had to stand to fire over the propeller arc. The added drag from this reduced the aircraft's already modest performance and few were used.

Above: The Dufaux C1 was another approach to designing a fighter without the use of a synchronizer. It had a buried engine and propeller in the middle of the fuselage, and was actually evaluated in combat.

Above: The Weymann W.1 of 1915 was another French attempt to design a successful fighter without a synchronized gun. Its 80 hp Clerget was mounted in the center of the fuselage, which caused serious cooling problems. Its structure was made of metal and it was to have two guns. It was abandoned in late 1915.

Above & Below: The F.E.2 served throughout the war. By 1917 the F.E.2d, powered by a 250 hp Rolls-Royce, could carry heavy amament albeit it was not very fast due to its pusher configuarion. As daylight combat became more intense it was used as a night bomber.

Above: The de Bruyère fighter was another French attempt to design an effective fighter without synchronizer. Apparently armed with a 37 mm cannon and possibly a machine gun, it was tested in April 1917, well after synchronizers were widely available. It was an innovative design powered by a Hispano-Suiza V8 in the fuselage driving an aft mounted propeller. Unfortunately, on its first test flight it rolled inverted immediately after takeoff and crashed. This was almost certainly due to its ailerons, which were pivoting wingtips. The pivot points appear to be too far aft, which caused the ailerons to go to full deflection at speed. With conventional ailerons it might have been a success. After the crash, it was abandoned.

Synchronized Machine Guns and Alternatives

Although the machine-gun synchronized to fire through the propeller arc was patented in concept in several countries before the war, it was not developed into a working model until Anthony Fokker achieved it in mid-1915. This development enabled Fokker to create the first true production fighter, the Fokker E.I.

Until the Fokker E.I appeared at the front, lack of a practical synchronization gear caused many work-arounds, many of them eccentric and dangerous. Roland Garros installed metal plates on the propeller of his aircraft to deflect bullets that hit it. Although there was danger from ricochets and perhaps breaking the engine mounts, potentially with fatal results, other French pilots also used deflector plates.

When Garros was forced down behind German lines, his propeller was rushed to *Idflieg* (German aviation inspectorate) in Berlin where its significance was immediately realized. *Idflieg* asked Tony Fokker to duplicate it and mount it on one of his airplanes. Fokker, seeing the obvious problems with this primitive device, offered a better solution; within 48 hours he returned with a machine gun synchronized to fire through the propeller arc without striking the blades. Although Fokker claimed in his autobiography that he solved the problem in this amazingly short time; photographic evidence shows that he had been working on the problem for some weeks.

French aircraft such as the Spad S.A.series and the Dufaux C1 had the engine and propeller in the middle of the fuselage, allowing a clear field of fire forward, a configuration also tested by the Royal Aircraft Factory B.E.9. The Dufaux

Above: Morane-Saulnier N with un-synchronized machine gun mounted; the metal deflector plate on the propeller blade is clearly shown.

Below: The metal deflector plate mounted on a Morane-Saulnier propeller blade.

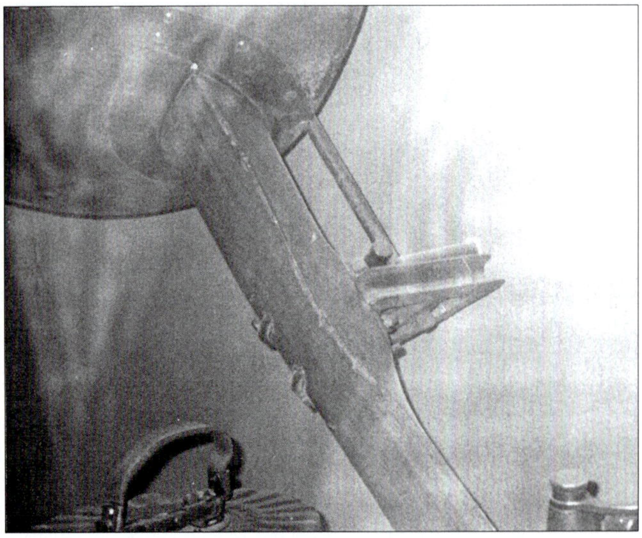

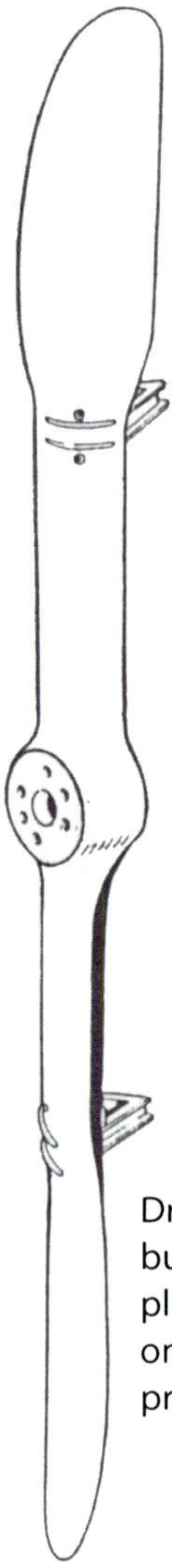

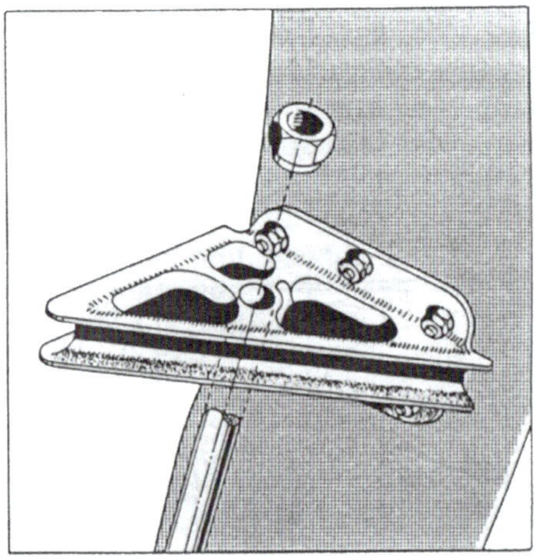

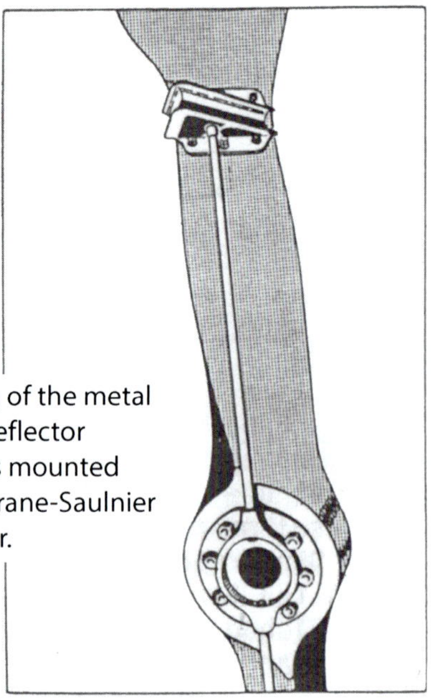

Drawing of the metal bullet-deflector plates as mounted on a Morane-Saulnier propeller.

Above: Jules Vedrines in the cockpit of his Morane-Saulnier N before a sortie; the metal deflector wedge on the propeller is clearly seen together with the un-synchronized Hotchkiss machine gun.

Above: Jean Navarre in the cockpit of his Morane-Saulnier N. The aircraft was fitted with an un-synchronized Hotchkiss machine gun with metal deflector wedges on the propeller. The M-S N debuted in June 1914, just before the war.

Above: The de Havilland DH.2 pusher was Britain's key contribution to the second generation of fighter aircraft. The prototype flew in July 1915, just as the Fokker Scourge was getting underway, and the design owed nothing to the subsequent air combat experience – the DH.2 initially featured a flexible gun like a two-seater, and the gun was only mounted in a fixed installation later. Most were powered by a 100 hp Gnôme Monosoupape engine; their Fokker E.II and E.III opponents also had 100 hp rotary engines. Its high-drag pusher configuration precluded future development.

Above: Jean Navarre in a Nieuport 16. Using the N.11 airframe, the N.16 was fitted with the heavier and more powerful 110 hp Le Rhône instead of the 80 hp Le Rhône used in the N.11. This made the N.16 faster and gave it a better climb rate at the expense of poorer handling qualities due to being somewhat nose heavy. The tractor N.11 was developed into an entire family of figher aircraft with higher power and better armament. The over-wing gun was a work-around due to lack of a functioning synchronizer at the time.

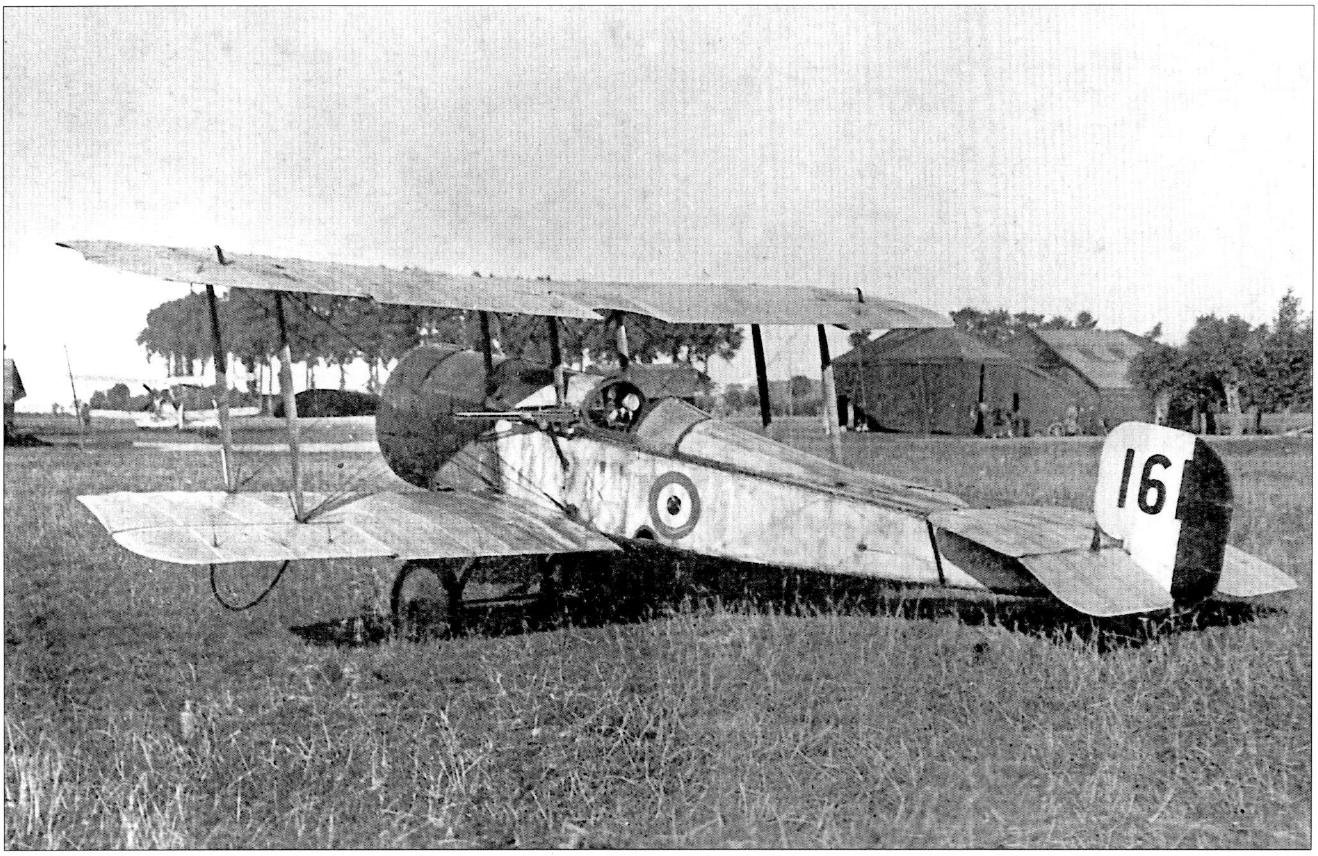

Above: The Bristol Scout was designed as an un-armed, high-speed single-seater for scouting. Later some aircraft were fitted with armament. Lack of a synchronizer meant Scout 1611 had to avoid firing through the propeller arc, and an oblique-firing gun was fitted. Major Lanoe Hawker downed three German aircraft with this scout to become the first fighter pilot awarded the Victoria Cross, Britain's highest award for valor in combat. Hawker's tactics were simple; "Attack Everything."

C1 was evaluated at the front and about 100 examples of the sturdier Spad S.A.series were built and used in combat by France and Russia. In French service only one German aircraft was confirmed shot down by a Spad S.A, and a Brandenbug C.I and one or two other aircraft were credited to an S.A.in Russian service.

The British Mann-Grimmer M1 had the engine in front but drove two propellers mounted between the wings to allow a gunner to fire forward. The Deperdussin TT used a much simpler solution; the gun was mounted on a framework that allowed the gunner, who had to stand up to fire the gun, to fire over the propeller. The problem with the TT was that the aircraft had to carry two crewmen, and the gunner generated excessive drag when he stood up to use his gun, reducing the aircraft's speed at a critical moment.

The most practical of these 'work-arounds' was mounting the gun over the upper wing, where it could fire forward over the propeller arc. Applying this to the Nieuport airframe resulted in a number of effective fighter designs that were widely used and ended the superiority of the Fokker E-type fighters. The British also used this approach, including with the well-known S.E.5a that combined an over-wing gun with a single synchronized gun. The Austrians also used this approach due to a shortage of synchronizing mechanisms, as did Belgium and Italy.

The 'Comic' version of the Sopwith Camel was designed as a night fighter to counter German Zeppelins and night bombers. Although the Camel was normally fitted with two synchronized machine-guns directly in front of the pilot, the muzzle-flash from these guns destroyed the pilots' night vision, so these synchronized guns were replaced by Lewis guns mounted over the wing where their muzzles were not visible to the pilot.

The French Spad 12 fighter and Spad 14 floatplane fighter were fitted with one synchronized machine gun and a single-shot 37mm cannon that the pilot had to hand-load in flight. Although a hit from a 37mm cannon was devastating, only expert pilots could hit a maneuvering target with a single shot. The French also fitted an experimental Nieuport monoplane fighter with a single-shot 'shotgun' mounted on the propeller hub.

Spad S.A.Prototype

Above: The Spad S.A.prototype at Buc in May–June 1915. The crew is test pilot Maurice Prévost and possibly Capitaine Pierra. (Musée de l'Air)

At the start of WWI the SPAD company began to make the Caudron G.3 and then the G.4 under license. Early on the chief designer, Louis Béchereau, realized a fighting airplane would need a forward-firing gun because "a hunting animal does not get his dinner by running away from it."

Béchereau had been chief designer for Deperdussin, so he modified a Deperdussin design into the TT, a two-seater with gunner in front. Unfortunately, the gunner had to stand to fire the gun over the propeller arc, creating a lot of drag and limiting performance. The engine vibration also made aiming difficult.

So Béchereau choose a different configuration when designing a fighting aircraft from scratch. There was no synchronizing gear for machine guns available when the SPAD S.A.was designed, so that limited Béchereau's choices.

Béchereau chose to place the gunner in front of the propeller in his own nacelle, or "pulpit" as it was popularly called. This gave him an excellent field of fire and of view like that of gunners in pusher aircraft, but with lower drag since the "lattice" structure supporting the tail was replaced by a more streamlined fuselage. Béchereau was granted patent number 22.088 for this design. The configuration required the engine and propeller within the fuselage and a relatively complicated support structure for the nacelle.

Both monoplane and biplane designs were contemplated, but the monoplane was abandoned in favor of the biplane. A second patent, 022.088, was filed on 7 June 1915, two weeks after the first flight of the S.A.prototype. This patent addressed the nacelle support system.

Béchereau submitted a third patent, 0488.191, on 4 June, 1915, describing a new system of interplane bracing for biplanes. This was incorporated into the S.A.prototype. This allowed a long-span wing with two bays of bracing be rigged as a single-bay wing; this reduced the amount of bracing cable exposed to the airstream, reducing drag.

Béchereau incorporated two features in the SPAD A that were used in all subsequent wartime SPADs. One was the hinging of the landing gear half-axle so that it moved up and down to more neatly track the forces encountered when the wheel contacted the ground while moving down and forward. The second was the aileron control system, also used on subsequent SPADs.

The S.A.prototype was built in the spring of 1915 and the first flight was on 21 May 1915. Because of the interest expressed by the Army, it was immediately presented to the SFA (Service de Fabrication Aéronautique) at Meudon, where it was tested by the Commission d'Examen des Avions et Moteurs. Lieutenant Toussaint, secretary of the Board of Examiners, wrote a short report on 29 May 1915 and submitted more complete reports on 2 and 3 June 1915. Extracts from the two reports are given in the table on the following spread.

The report of 2 June 1915 ended in remarks concerning the good and bad points of the aircraft divided into five categories.

Above: The Spad S.A.prototype at Buc in May–June 1915. The shapes of the major components are clear. (Musée de l'Air)
Below: The Spad S.A.prototype at Buc in May–June 1915. No gun mount is installed; the officer is the nacelle is armed with a carbine. (Musée de l'Air)

Above: The Spad S.A.prototype at Buc in May–June 1915. The observer does not appear to be armed. (Musée de l'Air)

Prototype S.A.Specifications	
Characteristic	
Wingspan	9.10 m (29 ft. 10 inches)
Length	7.30 m (23 ft. 11 inches)
Height	2.60 m (8 ft. 6 inches)
Wing Area	23.85 m^2 (256.5 $ft.^2$)
Tailplane Area	1.5 m^2 (16.1 ft^2)
Elevator Area	1.25 m^2 (13.5 ft^2)
Rudder Area	0.6 m^2 (6.5 ft^2)
Engine	80 hp Le Rhône
Propeller	Chauvière of 2.55m (100.4 in.) diam., 2m pitch
Airframe Wt.	237 kg (522.5 lb)
Engine, prop, hub	140 kg (308.6 lb)
Fuel & oil	95 kg (209.4 lb)
Fuel & oil tanks	23 kg (50.7 lb)
Payload	180 kg (396.8 lb) (pilot, gunner, gun & ammo)
Gross Wt.	675 kg (1488.1 lb(

Advantages

1. The field of fire and observation of the gunner is nearly unobstructed.
2. According to the manufacturer it is possible to replace the gunner with a special armament: for example, several machine guns that could be controlled by the pilot with a line-of-sight [perhaps a gun sight] that makes it possible for the pilot to aim, etc.

Disadvantages

1. The nacelle could obstruct cooling, although there was no overheating experienced during the 55-minute flight conducted on 31 May 1915. [This became a problem in operational service.]
2. The placement of the gunner could cause apprehension in the men assigned to that position due to the danger of serious injury if the aircraft is overturned on landing. The attachment of the nacelle to the propeller shaft is also criticized because with use it could possibly experience severe vibration. Currently there are very marked vibrations in the rear of the gunner. [Passage underlined in the report.] According to the manufacturer it would be possible to replace this configuration with another incorporating additional bracing to the upper wing. This would give greater safety but would significantly increase drag.

Prototype S.A.Flight Test Results			
Characteristic	**Test 1**	**Test 2**	**Test 3**
Fuel: 68.5 kg, Oil 24 kg		As test 1	As test 1
Payload	74 kg (pilot Maurice Prévost, observer 82 kg (unknown)	As test 1	74 kg (pilot Maurice Prévost, observer 82 kg (Lt Ardouin-Dumazet)
Rate of climb			
500 m	2 min 10 sec	2 mn 5 sec	2 min 50 sec
1000 m	4 min. 33 sec	5 min 45 sec	6 min 40 sec
1500 m	7 min 40 sec	9 min 30 sec	
2000 m	11 min 35 sec	—	16 min 0 sec
2500 m	16 min 5 sec	20 min 30 sec	21 min 45 sec
3000 m	21 min 5 sec	—	33 min 0 sec
3270 m	—	—	50 min 0 sec
Top Speed	—	150 km/h at ground level	—
	—	142 km/h at 2,100 m	—
	—	137 km/h at 3250 m	—
Stall Speed	—	89 km/h at ground level	—
On Test 3 the maximum ceiling "seemed to be" 3400 m.			

Notes:

- The aircraft tested was not provided with a machine gun. It would be interesting to carry out tests under real conditions, and to require that all fighting aircraft be presented with a machine gun, or a model of one, and the required mounting structure.
- At the present time there does not exist for the gunner a means of access to the nacelle.
- The visibility for the pilot could be improved.
- According to the tests, the aircraft presented by SPAD meets the conditions imposed for fighting machines except that the time to climb to 2,000 m was 11 minutes 35 seconds instead of the required 10 minutes.
- The large wing area is a good quality of the airplane. The wing loading is 28 kg/m2 (5.7 lb per square foot). This figure does not appreciably exceed the figures for the Caudron G.3 and Morane Parasol. From that point of view the airplane presents a certain safety factor, which a static test could confirm.

On 3 June 1915 Lt.Toussaint wrote a supplementary report concerning two climbing tests carried out at Meudon on 2 and 3 June 1915.

The prototypes wing bracing was described in patent number 0488.191 and the detail design was described in the Manual d'Aviation de la Première Guerre Mondiale, Tome II. The vertical members in the middle of the span were not true struts; they were intended only to act as vibration dampers and to limit deflection of the wings at mid-span. The complicated bracing in the cabane structure was not as effective as the designers expected, so production aircraft had diagonal bracing between the front cabane struts. Early photos of the prototype show no bracing between the upper wing and nacelle and it was probably flown that way at least once. Cables were added from the upper wing to the nacelle but it is not known when this was done.

Disposition of the Prototype

The prototype was given the SFA number S-11 and returned to the manufacturer. SPAD retained it for tests for an unknown time before returning it to the military. Photographed at Buc in March 1916, the aircraft was almost completely modified into the standard S.A.1 configuration; the difference is that the original horizontal tail surfaces had been retained. There also appears to be only a single bracing strut between the tailplane and the fuselage and no national markings.

Above & Below: The Spad S.A.prototype in flight at Buc in May–June 1915. The crew is test pilot Maurice Prévost and possibly Capitaine Pierra. (Upper photo: SHAA, lower photo: Musée de l'Air)

Above & Below: The Spad S.A.prototype at Buc in May–June 1915 during testing. The crew is test pilot Maurice Prévost and possibly Capitaine Pierra. (Upper photo: SHAA, lower photo: Musée de l'Air)

Above: Spad S.A.prototype fitted with flight test probes. In this early configuration there are no bracing cables from the wing to the gunner's nacelle. (Musée de l'Air)

Above: The Spad S.A.prototype at Buc in May–June 1915. There are no bracing cables from the upper wing to the nacelle, but there are windshields for both pilot and gunner. (Musée de l'Air)

This close-up of the Spad S.A.prototype shows the wicker cooling grill. There is no diagonal cabane bracing or bracing cables from the wings to the nacelle. There are no entry step or gun mount. On the left is SPAD's test pilot Maurice Prévost. (Musée de l'Air)

Above: The Spad S.A. prototype with standard gunner's nacelle showing how the nacelle folded forward for engine maintenance. The engine is an 80 hp Le Rhône rotary. (Musée de l'Air)

Above: The Spad S.A. prototype with a mock-up nacelle installed. This appears to be a differnt nacelle than the photo above. (Musée de l'Air)

Above: The Spad S.A. prototype fitted with a pulpit without gunner's position. The cooling vents have been enlarged also. (Jean-Amigues)

Above: The Spad S.A. prototype in March 1916. The horizontal stabilizers are original, but otherwise the aircraft has been modified into the production S.A.1 configuration. (Musée de l'Air)

Above: Closeup of the Spad A prototype with crewmen.

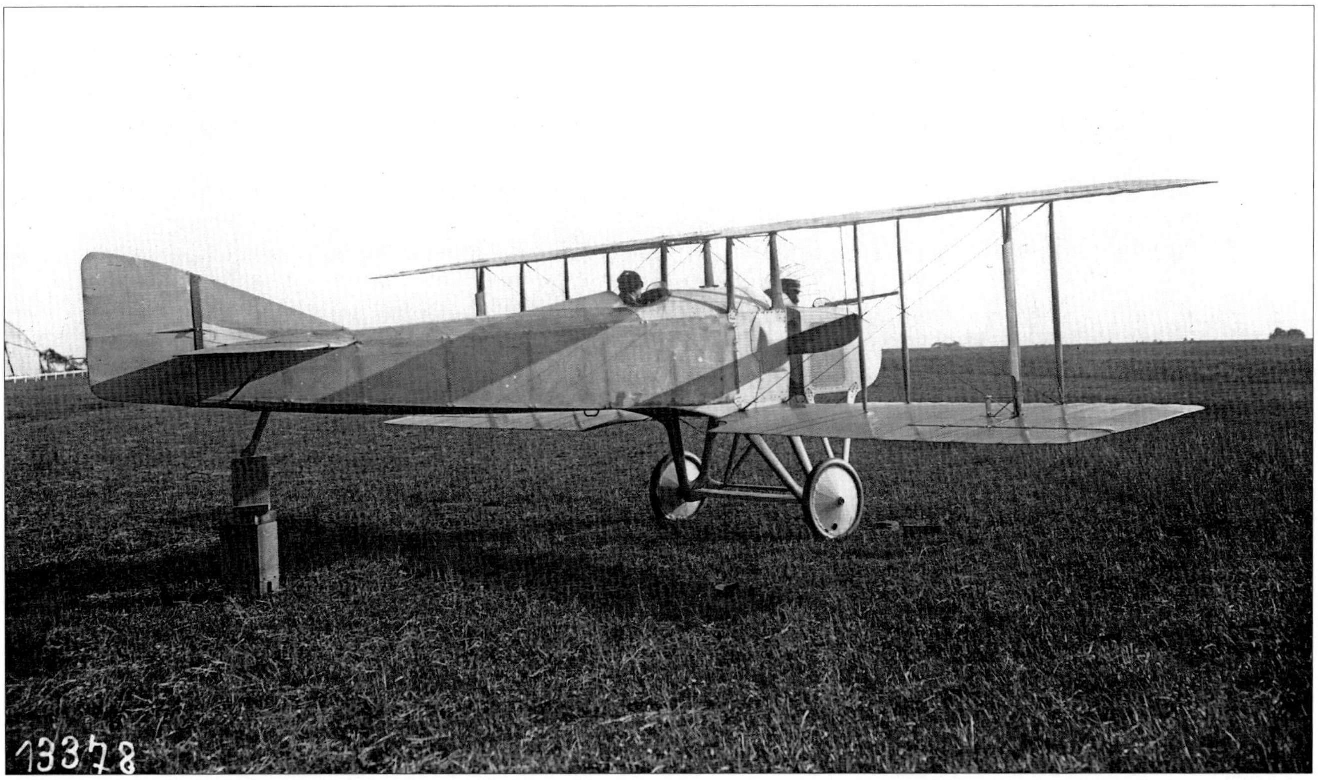

Above: The Spad A prototype with crewmen.

Above: The Spad S.A.prototype is the center of a lot of attention. (Musée de l'Air)

Spad S.A.1

Above: S-1 was the first production example of the Spad S.A.1 but was not quite the final design. The steel fitting at the top of the pylon differs from the production standard and the three holes at the front of the fuselage seen on later production S.A.1 aircraft are not present. The cover over the engine resembles the prototype more than the production S.A.1. This aircraft was sent to MS 23 at Châlons-en-Champagne in the summer of 1915. The pilot is Capitaine Maurice Schlumberger and the gunner is Capitaine Max Bousche. (SHAA)

The SPAD S.A.1 was the first new two-seat fighter brought into service in the Aviation Militaire Française (French Military Aviation); all preceding two-seat fighters were adaptations of existing aircraft.

Following the tests carried out on the prototype in May and June 1915, on 5 June the Ministry of War directed the commandant the SFA at Meudon to buy the prototype and place an order for 10 type SA.1 production aircraft from the SPAD company.

The Ministerial Order Stated:

Because of the tests carried out by the type S.A.L. (sic), the heavy artillery section with the 80 hp Le Rhône and because of the potential which it will present as a fighter, I decide to put in an order for ten of this type. Moreover, you will proceed to purchase the test machine, which will be issued to the General Reserve where it will accomplish a flight of three hours endurance. I draw your attention the advantage there would be to be able to provide the 10 aircraft referred to above with 110 hp Le Rhône instead of the 80 hp Le Rhône.

There are several items to note:

1. The aircraft is described in the text as being of type S.A.L. This name is found in other period documents and requires an explanation. It is a result of a typographical error due to the use of the lower case letter "l" on typewriters of the time to stand for the number "1". Because of this similarity someone transformed the "l" into "L". This error was corrected almost immediately in later reports as of 12 June 1915.
2. This order required the prototype of carry out a flight of 3 hours endurance with the Réserve Générale Aéronautique (Aeronautical General Réserve, R.G.A.). This was the wish expressed by Lt. Toussaint, secretary of the Centre d'Essais Aériens Militaires (Military Aviation Test Center) on 29 May 1915. The test was intended to determine whether the cooling of the 80 hp Le Rhône was adequate in spite of the restricted cooling airflow due to the nacelle. (In retrospect the cooling would never be satisfactory for any SA type.)
3. As early as the first order for the machine, the French military wanted to install the more powerful 110 hp Le Rhône 9J. This only became available in production

Above: Capitaine Maurice Schlumberger posing in front of S-1. He had a great deal to do with the development of the S.A.1. Unfortunately, he was KIA in S.A.1 S-10 near Aure on 22 November 1915. (SHAA)

S.A.1 S-1. (SHAA)

Above: S.A.1 S-1. (SHAA)

Below: S.A.1 S-1. (SHAA)

Above: S.A.1 S-1. (SHAA)

Right: S.A.1 S-1. (SHAA)

quantity a few months later for installation on the SA.2.

A 30 June 1915 War Ministry message to G.Q.G. Aeronautics notes the fate of the prototype: The Béchereau S.P.A.D. test aircraft will be delivered to Buc next 2 July and assigned to the R.G.A. after reception if no new modification is necessary.

The prototype was photographed at Buc in March 1916, by which time it had been modified into the S.A.1 configuration, except that the triangular horizontal tailplane was retained. It was assigned the SFA number S-11, but at the time of the photograph the number had not been painted on the rudder or longerons. Unanswered questions include: Why was the triangular tailplane retained when so much effort was spent changing everything else? At what time was the SFA number assigned? Was the SFA number ever applied to the prescribed locations on the aircraft? What was the final fate of the aircraft?

The British RFC, which had only the Vickers FB-5 Gunbus at the time, very early on expressed an interest in the SPAD S.A.Two RFC officers, Major S. Lewis (commander of No.3 Squadron) and Captain T. O'Brian, examined a dismantled SA and submitted the following report on 2 August 1915:

Report on the New SPAD Biplane

This was dismantled. It appears to be a well-designed machine.

The passenger sits in a nacelle fixed in front of the propeller. This gives a perfect field of fire and view.

It is about the same speed as the Morane biplane with the same engine, probably rather faster.

It probably lands faster and is harder to get into a smaller field.

It would be expensive in observers if piloted by indifferent pilots. A French flying officer told me that it was easier to fly than either a Morane Parasol or a Nieuport.

I consider that it should be tried when re-assembled.

Instead of a passenger, five or six machine guns can be fixed in the nacelle and fired from the pilot's seat.

S.A.1 Specifications

Characteristic	
Wingspan, Upper	9.100 m (29 ft. 11 inches)
Chord	1.544 m (5 ft.)
Length	7.300 m (23 ft. 11 inches)
Height	2.600 m (8 ft. 6 inches)
Wing Area	25.35 m^2 (276.9 ft.^{2})
Empty Wt.	425 kg (937.0 lb)
Useful Load	300 kg (661.4 lb.)
Total Wt	710 kg (1,565.3 lb.)
Engine	80 hp Le Rhône 9C rotary
Propeller	2.55 m (8 ft. 4 in) diam. Chauvière
Ceiling	4,300 m (14,108 ft.)
Endurance	2 hr. 45 min.
Top Speed	135 km.h (83.4 mph) at sea level
Climb Rate	
1000 m	6 min. 30 sed.
2000 m	12 min. 30 sec.
3000 m	23 min. 30 sec.

Aft view of S-1. (SHAA)

Above: Cadet pilot Jacques Roques at the School of Aviation at Pau photographed in S-3 in August 1916, where is was being used as a trainer. The aircraft appears well used since the fabric looks ragged. (SHAA)

Above: S.A.1 with ground crew. (SHAA)

Above: SFA number S-4, a production S.A.1.

Above: Probably S-4, a Spad S.A.1 that has apparently been modified with larger side ducts and a double-tube machine gun mount. (SHAA)

Above: SA.1 S-4 undergoing maintenance. (SHAA)

Above: SFA number S-5 a production SA.1.

Above: Spad S.A.1 S-6. (Jean-Amigues)

Above: Spad S.A.1 S-6. (Jean-Amigues)

Above: S-7, a production Spad S.A.1, being delivered at Buc in September 1915 (SHAA)

Above: S-7, a production Spad S.A.1, being delivered at Buc in September 1915. (SHAA)

Above: S--7 next to a single-seat Nieuport 10 (with 110 hp special) of Sergeant Georges Guynener (probably N436) with over-wing gun at N 3, winter 1915–1916. Note there is no incidence bracing brtween the interplane false struts of S-9. (SHAA)

Above: Production Spad S.A.1 showing some oil stains on the front of the fuselage and a window under the forward part of the nacelle. This is S-9 of Capitaine Brocard, commander of Esacdrille N3 at Breuil-le-Sec at the end of 1915. (Musée de l'Air)

Above: This appears to be Capitaine Brocard's SA.1 S-9 with Guynemer's single-seat Nieuport 10 in the baciground. (SHAA)

Above: Spad S.A.1 S-9 from Escadrille N-3 nest to the Nieuport 11 prototype (N-402). (SHAA)

Above: S--9 of Capitaine Brocard next to a single-seat Nieuport 10 (with 110 hp special) of Sergeant Georges Guynener (probably N436) with over-wing gun at N 3, winter 1915–1916. Note there is no incidence bracing brtween the interplane false struts of S-9. (SHAA)

Above: Spad S.A.1 S-9 from Escadrille N-3 next to the Nieuport 11 prototype (N-402). (SHAA)

Above: S--9 of Capitaine Brocard next to a single-seat Nieuport 10 (with 110 hp special), winter 1915–1916. (SHAA)

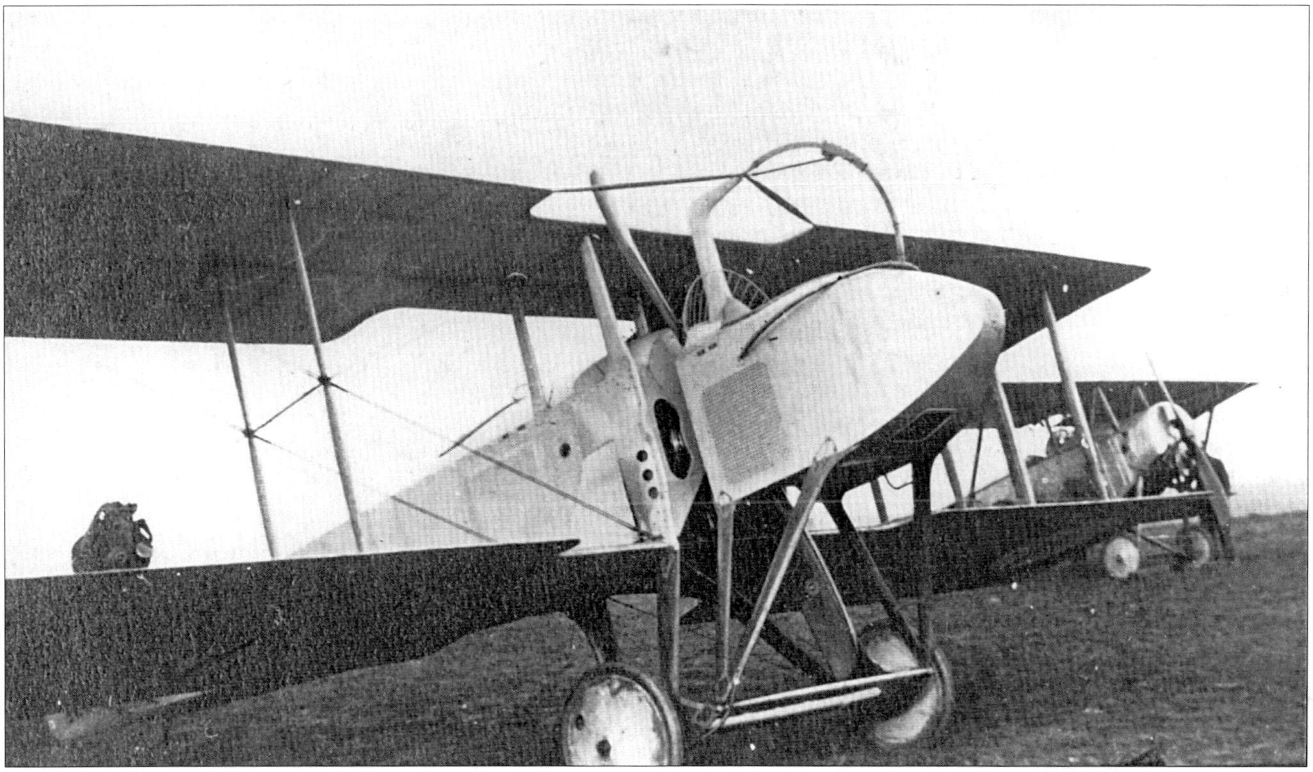

Above: S--9 of Capitaine Brocard next to a single-seat Nieuport 10 (with 110 hp special) of Sergeant Georges Guynener (probably N436) with over-wing gun at N 3, winter 1915–1916. Note there is no incidence bracing brtween the interplane false struts of S-9. (SHAA)

Above: S--9 of Capitaine Brocard next to a single-seat Nieuport 10 (with 110 hp special) of Sergeant Georges Guynener (probably N436) with over-wing gun at N 3, winter 1915–1916. Note there is no incidence bracing brtween the interplane false struts of S-9. (SHAA)

Above: S.A.1 S-9 from escadrille N-3, wearing late markings – a little tricolor flag on the tail, and a cockade on nacelle. (SHAA)

Above: S-9 from escadrille N-3, during winter 1915-1916. . (SHAA)

Above & Following Page, Above: Probably S-10, a production Spad S.A.1, of Captiaine Maurice Schlumberger, the tall officer in the center wearing a kepi. Above, French soldiers removing things out of the left baggage compartment. Below and following page, the gun mounts is pointed to the right, indicating the centering springs were overridden. This aircraft was assigned to N 23 at the end of 1915. The SFA number of the rudder has been painted over. The diamond marking on the nacelle is red, white, and blue, the blue being a center diamond washed out in these images. (SHAA)

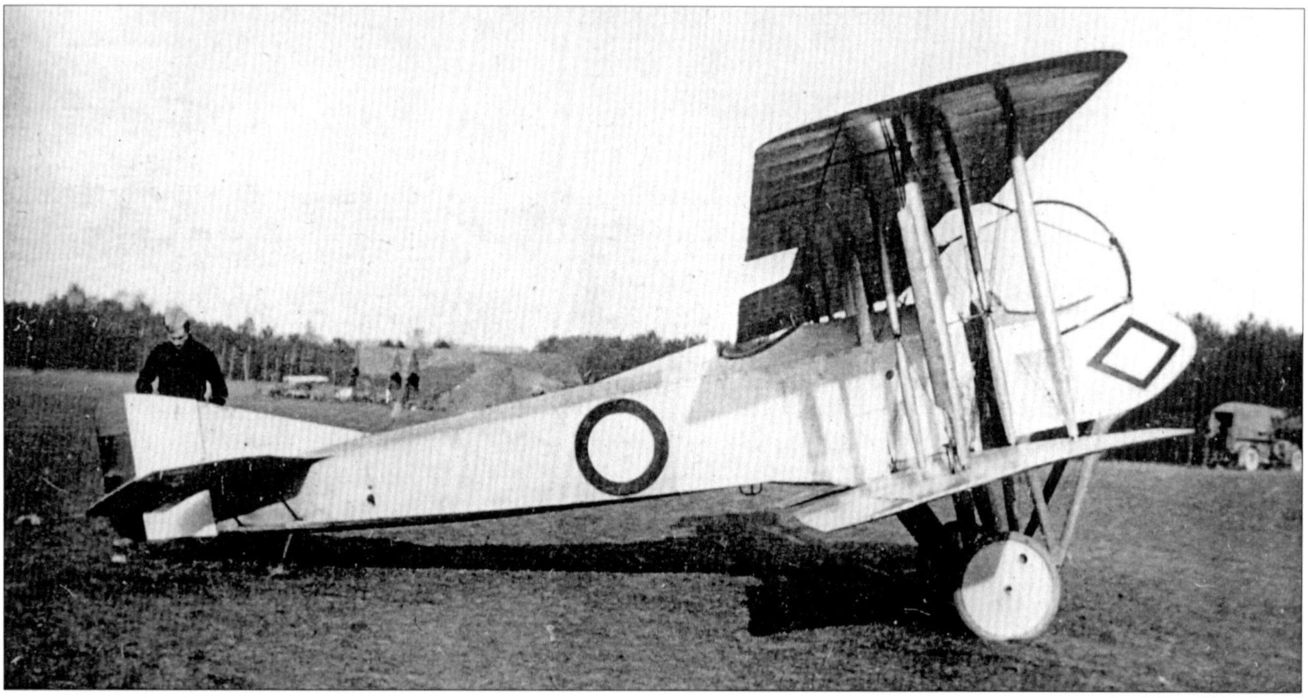

It cannot be flown without the equivalent of a passenger's weight in front.

Did the officers mistake a Rhône 9C for a 9J, or was there actually a 110 hp Rhône installed? The 1915 model of the 110 hp Le Rhône, which had a smaller diameter than the eventual 9J, was being tested in August 1915 in Moranes and Nieuports, but it tended to routinely burn its valves and was abandoned. Apparently there was an S.A.2 with the 1915 model engine received around 20 October 1915, but it was not sent to the front until it had a 9J installed. Because the British officers looked at a dismantled aircraft, it may have been the S.A.2 that had the early model engine and after testing it was dismantled for repairs, which is reasonable considering the history of the early engine. How far the dismantling had progressed was not mentioned; the engine may have been removed. In any case a later, undated, anonymous British report was less than complimentary:

In this machine the passenger is slung in a small fuselage in front of a tractor propeller. This arrangement is considered to be unnecessarily dangerous and it is not justified by the objects attained as regards arc of fire.

All the Spad machines are of similar type and are considered to be of no interest in their present form.

In his book "*One Man's War*", Bert Hall describes a macabre episode at Avord on 11 October 1915 at Escadrille N68:

The first Spads were powered by an 80 hp Rhône engine and had a little wicker nacelle out in front of the actual body of the machine where the observer sat and did his fighting. The pilot sat back of the top wing, the ship being a biplane.

The front nacelle was attached to the top wing and to the landing carriage. The first one they tried turned over taking off and mashed the observer flat as a flapjack. Everyone on the field tried to argue them out of trying the second one, but nothing would do but a test flight. So a second one was hauled out and two Frenchmen hopped it. It was their last hop. The pilot was a friend of mine, and I regretted his death a great deal. His name was Lieutenant Stromeyer. The machine took off all right but when Stromeyer tried to pull it out of a gradual dive, the front nacelle came off and the wings buckled up. And that was the end of the first model Spad. The design was washed out and a new one was developed. (Stromeyer survived, gunner Soldat Michel was killed.)

The first SFA number S.A.1 was S-1 and it was sent to Escadrille MS 23 at Châlons-en-Champagne during the summer of 1915. Photos show an aircraft that is almost but not quite the final configuration. A steel fitting at the top of the inverted "L" gun mount is not seen on later S.A.1 aircraft starting with S-3. No photo is available of S-2.

The prototype had no diagonal bracing between the front cabane struts but S-1 had these cables. Photos of S-3 show tie rods between mid-points of the false struts between the wings. The first S.A.2, S.12, appears to be the first aircraft to have incidence bracing between the false struts. On S-1 the front of the fuselage and side ducts were different shape and the three holes for exhaust and oil seen on later production aircraft are missing from the forward fuselage.

Delivery of the ten S.A.1 aircraft got underway in September 1915 when they started to arrive at Buc. They joined the R.G.A. about October and were then delivered to some frontline escadrilles. Only MS 23 can be positively identified. French fighter units were in the midst of standardizing on the Nieuport 10 and then the Nieuport 12 two-seaters and the single-seat Nieuport 11. The underpowered S.A.1 aircraft were soon replaced by the more powerful S.A.2s, and after frontline service were sent to serve out their lives at the flying schools.

This Page: The Spad SA.2 S-10 of Capitaine Maurice Schlumberger. This aircraft was assigned to N 23 at the end of 1915. The SFA number of the rudder has been painted over. The diamond marking on the nacelle was red, white, and blue, the center being blue. The center blue diamond is washed out in some photos but is visible in the photo at below right.

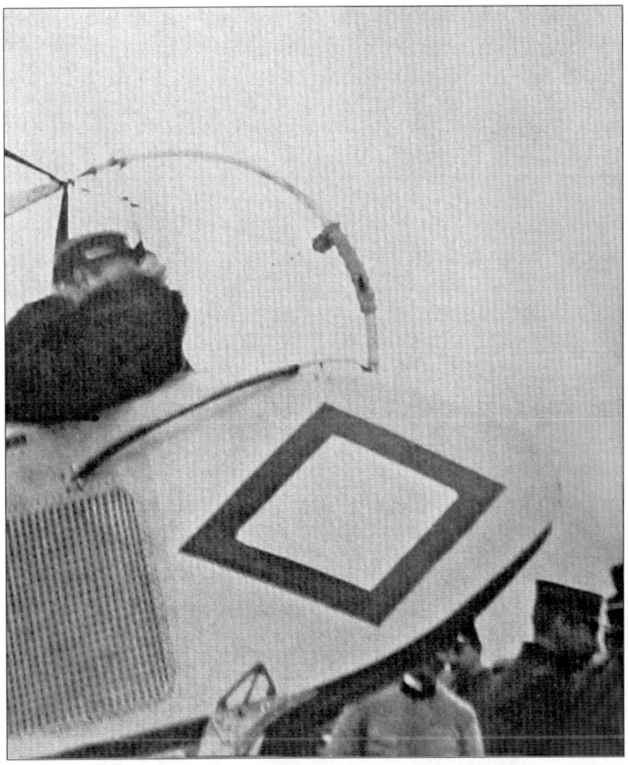

Above & Below: A Spad S.A.1of the Camp Retranché de Paris (C.R.P.) at Le Bourget-Dugny has just overturned, and the expressions on the men indicate the passenger may not have fared well. The lower panel and fabric have been peeled from the nacelle. Could this be one of the accidents that Bert Hall referred to? (Musée de l'Air)

Above & Below: Spad SA.1 after a bad landing. This was the dread of all Spad A observers, who were very vulnerable in this kind of accident.

Left: An S.A.1. (Jean-Amigues)

Above: An S.A.1 possibly photographed at the R.G.A. at Bourget-Dungny.

Above: Pilot posting with an S.A.1. The document attached to the left wing may indicate delivery of a new airplane, (SHAA)

An S.A.1 getting ready for a mission.

Above: S.A.1 being prepared for flight. (SHAA)

Above: S.A.1 from escadrille N-23 during January 1916. (David Méchin)

Above & Facing Page: The wreck of Capitaine Maurice Schlumberger's S.A.1 S-10, shot down on 22 November 1915. Unfortunately, Capitaine Schluberger was killed in this crash. Many construction details are visible. (SHAA)

Above: Pilots and observers of MS 23 during 1915. Then-Lieutenant Maurice Schlumberge is on the far right. (SHAA)

Spad S.A.2

Above: S-12 was the first production SA.2. The SA.2 adopted the more powerful 110 hp Le Rhône and revised cooling louvers. S-12 is shown after a red-white-blue chevron marking was applied. (SHAA)

Above: S-13 was the second production SA.2. The number '12' on the fuselage is the early marking of Escadrille 12. But S-13 on the tail is its serial number. The SA.2 adopted the more powerful 110 hp Le Rhône and revised cooling louvers. It is shown after a red-white-blue chevron marking was applied. (SHAA)

Above: Standard mid-production S.A.2 S-15 clearly showing the large cooling ducts. (SHAA)

From the beginning the French military authorities wanted more power and performance and the additional weight and drag of the gun and mounting reduced performance compared to the unarmed prototype, making additional power essential. Finally, availability of the more powerful 110 hp Le Rhône 9J enabled production of the more powerful SA.2. Visually, the SA.2 adopted larger cooling vents to help cool the more powerful engine.

A purchase order for 40 SPAD SA.1, SA.2, and SA.3 aircraft was dated 6 September 1915. Each aircraft was priced at Fr 17.000 for a total cost of Fr 680.000. The prototype had been purchased for Fr 28,000. The following is extracted from a dispatch send on 2 September 1915.

In the conference of the inspectors of aviation material for the armies on 20 August, the pilots at the front expressed the desire that the replacements of planes and engines of the three SPAD escadrilles to be formed would be sent to 2nd reserve of supply. This provision would make it possible to hasten the use of the planes while waiting for the delivery of the rolling stock (ground equipment) and the accessories other than the replacements.

The dispatch proves that French Military Aviation originally wanted to establish up to three escadrilles of fighters equipped with SPAD SA aircraft. The three escadrilles were to be S 67, S 68, and S 69. S 67 was to be formed on 10 October 1915, S 68 on 30 October, and S 69 formed in November. This did not happen because the military passed the SA.2 aircraft out in small numbers to various escadrilles as was done with the SA.1. The French may have wanted to test their operational abilities without crippling any one escadrille. However, superior aircraft were in the pipeline and the career of the SA series was short.

The September purchase order was for 29 SA.2 aircraft, ten SA.1 aircraft, and one SA.3 aircraft. With the modification of the prototype into an SA.1, that makes a total of 41 SPAD SA series in French service.

Characteristics

Installation of a more powerful engine in the SA airframe required some changes to the airframe. Because cooling was marginal in the SA.1 and would be an even worse in the more powerful SA.2, the SA.2 had larger cooling ducts and expanded cooling area in the front engine mount. The extra weight of the larger engine partially solved the tail heaviness of the SA.1. Photos indicate the larger cooling ducts were retrofitted to a few SA.1 aircraft. Also, there were SA.4 aircraft with and without the larger ducts. Additionally, a few SA.2 aircraft may have reverted to the 80 hp Le Rhône 9C. This is shown in a photo of S-46 SA.2 with PU 160/PC 100, which indicates the 89 hp engine.

The SA.2 used a modified gun mount with the addition of a second curved tube so the gun was now mounted between two tubes. Interestingly, S-4 was used to evaluate both the larger cooling ducts and the double-tube gun mount.

Frontline Service

Starting in December 1915 the SA.2 aircraft were dispensed in small numbers similarly to the SA.1 experience.

Above: Standard mid-production S.A.2 S-16. (SHAA)

Above: Standard mid-production S.A.2 S-16. (SHAA)

Above: S.A.2 S-16. (SHAA)

Above: S.A.2 S-16. A mid-production SA.2, it clearly shows the large cooling ducts and standard dual tube machine gun mount. (SHAA)

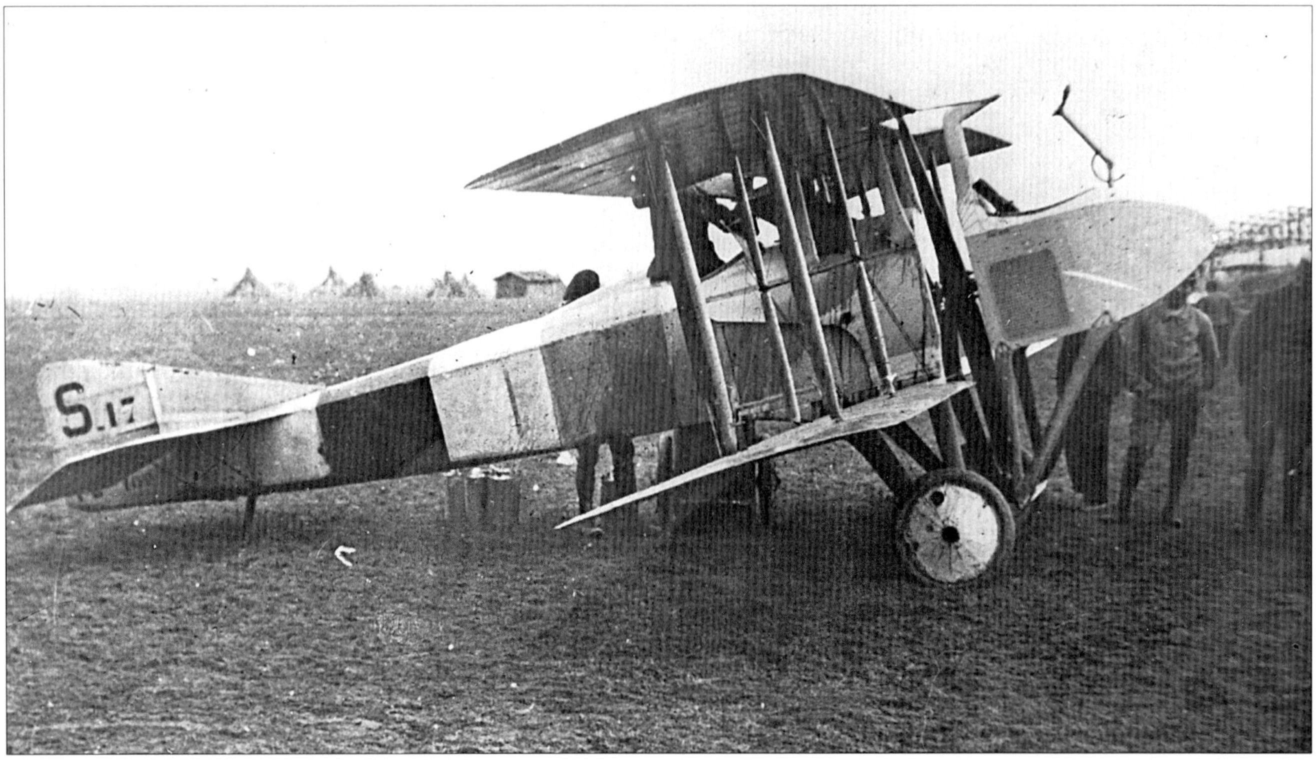

Above & Below: Two photos of S.A.2 S-17. A mid-production SA.2, it clearly shows the larger cooling ducts and standard machine gun mount. (SHAA)

Above: Another photo of S.A.2 S-17. A mid-production SA.2, it clearly shows the larger cooling ducts and standard machine gun mount. (SHAA)

Above: *MA JEANNE* ("My Jane" in English) was S.A.2 S-19. (SHAA)

Above & Below: More photos of *MA JEANNE*, S.A.2 S-19. A mid-production S.A.2, it clearly shows the larger cooling ducts and standard machine gun mount with dual metal tubes. (SHAA)

Deployment of the SA.2 was complete in March 1916. Only four "SPAD-Béchereau" were in frontline service and five were in reserve and training schools. This small number of aircraft at the front were insufficient to supply an escadrille and the idea of forming units specifically with SPAD SA aircraft was not pursued. The SA's poor performance definitely contributed to outcome.

The majority of the SAs were assigned to the Camp Retranché de Paris (C.R.P.), an assignment that was quiet compared to frontline units. The SA.2 was assigned at different times to Escadrilles N 3, N 23, N 49, N 73, and the C.R.P.

The SAs contemporaries were likewise primitive, as was the SA series. Aside from the nacelle supporting structure, the SA was otherwise sturdy and had good flying qualities. However, the nacelle configuration was forced on the SA due to lack of a synchronizing gear when it was designed and built. The result limited their operational service to about six months and kept production small.

On 17 May 1916, N 73 was notified of an incursion by a German reconnaissance aircraft. Caporal Robert Sabatier had no other aircraft so took the training machine, an S.A.2, SFA number S-31. He had no gunner so talked another pilot Caporal Claude Funck-Brentano, into going with him as gunner. Climbing for altitude they spotted three enemy aircraft on their way back to the German lines and immediately went on the attack. The crew of S-31 persisted attacking the enemy despite their aircraft being damaged.

S.A.2 Specifications	
Characteristic	
Wingspan, Upper	9.555 m (31 ft. 4 inches)
Wingspan, Lower	9.250 m (30 ft. 4 inches)
Chord	1.544 m (5 ft.)
Length	7.300 m (23 ft. 11 inches)
Height	2.600 m (8 ft. 6 inches)
Wing Area	23.36 m^2 (250.8 ft.^{2})
Empty Wt.	435 kg (959.0 lb)
Useful Load	300 kg (661.4 lb.)
Total Wt	735 kg (1,620.4 lb.)
Engine	110 hp Le Rhône 9J rotary
Ceiling	4,300 m (14,108 ft.)
Endurance	2 hr. 30 min.
Top Speed	154 km.h (95.7 mph) at sea level
Climb Rate	
1000 m	6 min. 30 sed.
2000 m	12 min. 30 sec.
3000 m	23 min. 30 sec.

Above: S.A.2 S-19 before it received the inscription "*MA JEANNE*". (SHAA)

Above: A mid-production SPAD A.2, very probably *"Mister Fly"*. (SHAA)

One of the three Germans crashed in French lines but the modest performance of the SA made it unable to catch the other aircraft. The SA crew brought their badly damaged aircraft back to their airfield where it was immediately dismantled. Both Frenchmen received the Médaille Militaire. In addition Funckh-Brentano was mentioned in dispatches and was promoted sergeant. He achieved one more confirmed victory before he was killed in aerial combat less than two months later.

As far as is known, this is the only victory of a SPAD SA in France.

Around the same time, Escadrille N.67, stationed at Froidos on the Verdun front, received a Spad A.2 that Jean Navarre, the ace nicknamed "the Sentinel of Verdun," wanted to try out, taking on board a young observer, Second Lieutenant Jacques Leps (a future ace with 12 victories), as a gunner. However, he was wounded almost immediately, and Leps flew with another pilot. Leps's account, preserved by the French Air Force Historical Service, is very informative about the aircraft's flight characteristics:

"On June 19, 1916, I took off for the first time in the Spad A.2 over the lines. We had a dogfight during which there were, naturally, many acrobatic maneuvers by both the Germans and the French. It was the first time the pilot, Sergeant Jules Moulias, had flown this aircraft. When this aircraft dived, it dived much faster than my body. I was leaving my seat, held only vaguely by my tethers and the makeshift roof of the gondola. Firing was impossible. On the contrary, at the pilot's aid, I was literally crushed in my seat: impossible to fire a single shot… The pilot finally realized I couldn't fire. He left the battle zone and returned to the Froidos airfield. The entire squadron, with Captain de Saint Sauveur at the head, was there to see the aircraft land and ask for our impressions. The pilot approached the ground as best he could, at over 100 kilometers per hour, and he flipped over. The pilot was in no danger. But me, in my small gondola, with the large rotary engine spinning behind me? I was literally crushed! The entire squadron rushed over. There were 20, 50 arms pulling me from the wreckage. I regained consciousness, and my first words were: "I don't want to be piloted anymore, I want to pilot!" The emotion was so intense that I was taken directly to another airfield. I never saw that Spad A.2 again…"

Leps quickly obtained his transfer to flight school… As for the Spad A.2s, he claims that they were grounded. In any case, they disappeared from the squadrons and were relegated to training duties after barely three months of use at the front.

Above: S.A.2 *Mister Fly* showing the largercooling ducts and machine gun mount. (SHAA)

Above: Standard mid-production S.A.2 clearly showing the large cooling ducts. (SHAA)

Above & Below: S.A.2 S-31 and crewman. (SHAA)

Above: S.A.2 S-31 when assigned to N 73. S-31 scored the only confirmed victory of an SA on the Western Front. (SHAA)

Above: This SPAD A.2 marked with the number 32 on its nacelle was used for training on the Buc airfield. (SHAA)

Right: S.A.2 S-39 with unit crewman. (SHAA)

Below: S.A.2 after a rough landing. It took a lot of courage for gunners to flying in the SA series knowing this was a not uncommon occurance. (SHΛΛ)

Close-up of an S.A.2 with the standard dual tube machine gun mount. (SHAA)

S.A.2 and crewman. (SHAA)

Above: S.A.2 at the front with crewmembers.

Above: On May 17, 1916, Sgt Théophile Funck-Brentano and Sgt Robert Sabatier pose on the return from their victory against an Aviatik C which crashed in the vicinity of Chatas (88) - During this fight, they were flying on board SPAD A2 No. 31 and although they were both qualified military pilots, it was Sabatier who was piloting. Their plane was damaged by enemy fire. (Jean Amigues)

Above: SPAD A.2 n°31 being repaired. (Jean Amigues)

Close-up of an S.A.2 with the standard dual tube machine gun mount. (SHAA)

Above: Unidentified Spad S.A.2. (Jean-Amigues)

Pilot familiarizing himself with unidentified Spad S.A.2. (Jean-Amigues)

Above: Unidentified S.A.2 at the front.

Right: Spad A.2 decorated with a two-color chevron, possibly blue and red, suggesting an airplane from Camp Retranché de Paris (CRP). (David Méchin)

SKETCHES FROM "MANUEL D'AVIATION de la PREMIER GUERRE MONDIALE: TOME II"

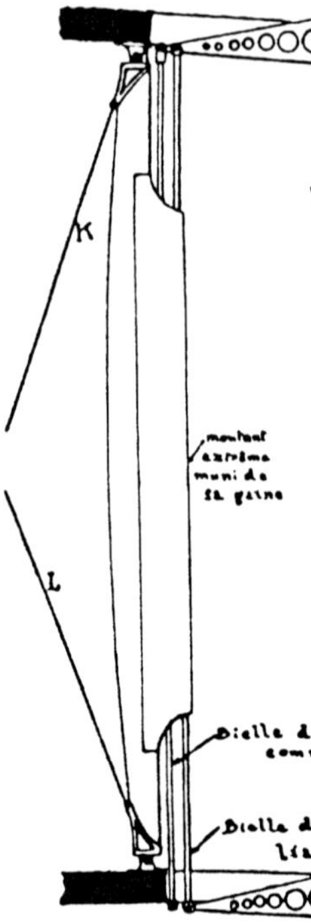

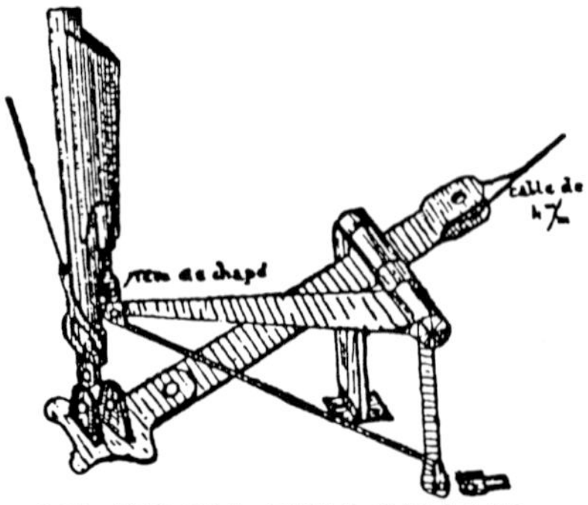

AILERON BELLCRANK

SA-2 AILERON CONTROL

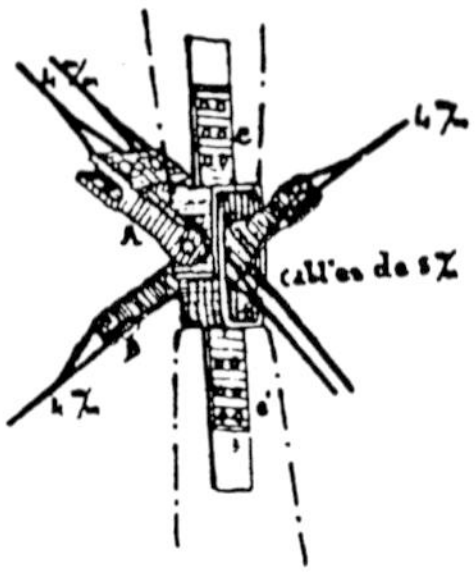

CENTRALJOINT INTERMEDIATE BAY

TAILSKID MOUNT STRUCTURE

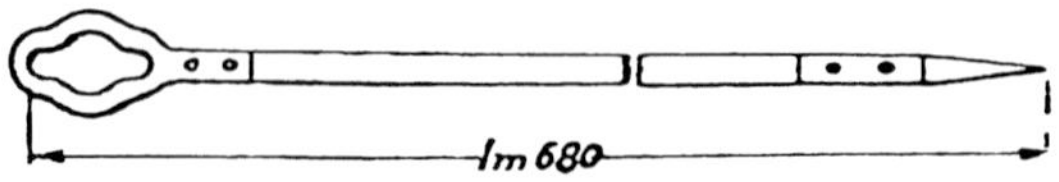

WING ALIGNMENT TOOL

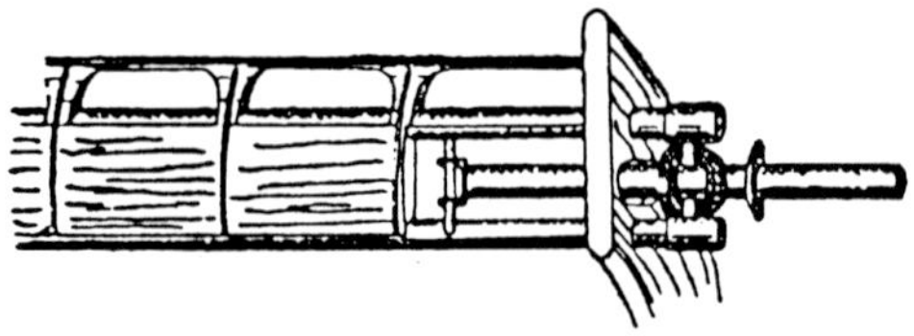

SPREADER BAR AND AXLE

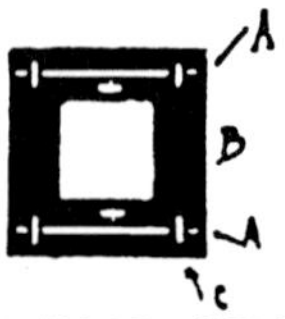

WING SPAR SECTION

CARBURATOR

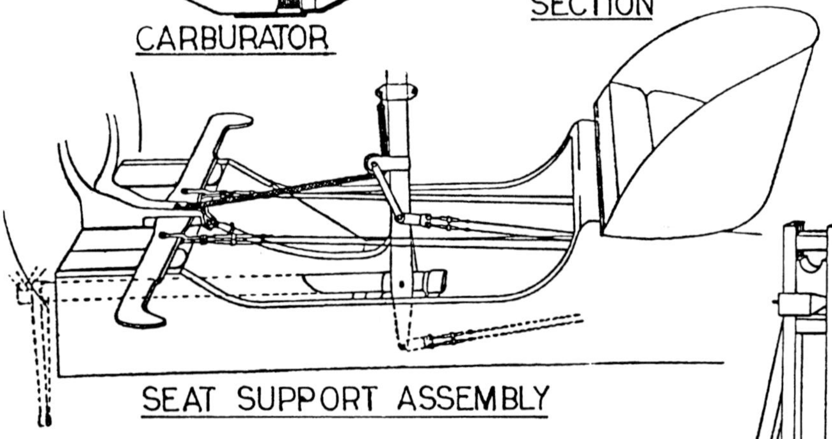

SEAT SUPPORT ASSEMBLY

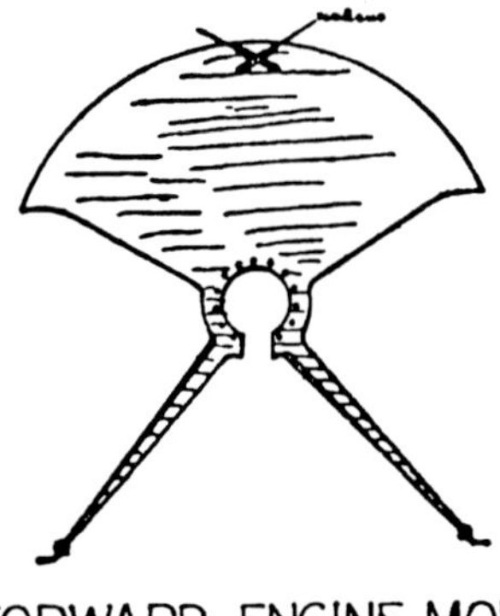

FORWARD ENGINE MOUNT

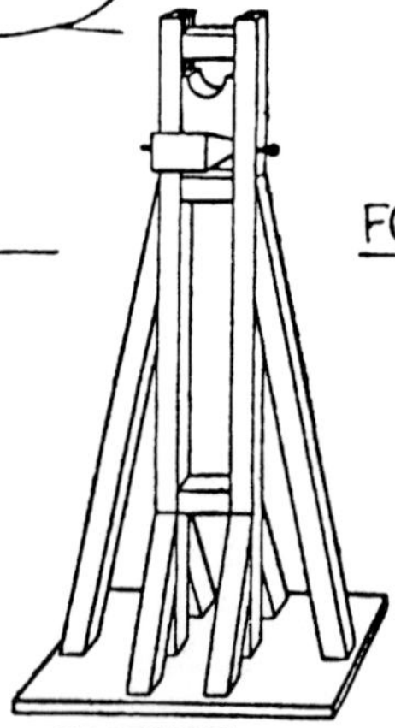

ENGINE REMOVAL STAND

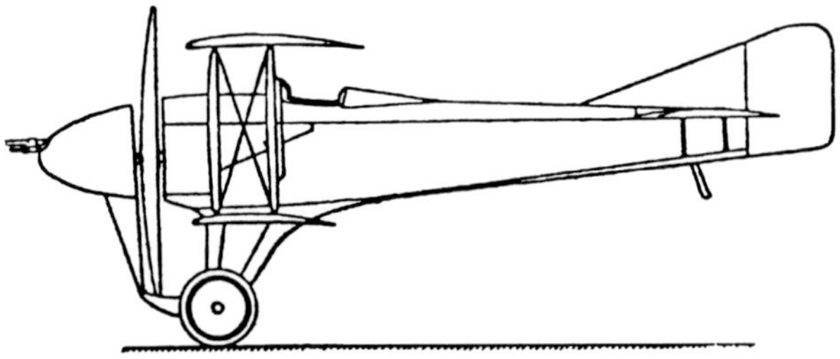

SG

SMALLER NACELLE, WINGS MOVED AFT, SOLID NACELLE SUPPORT STRUT

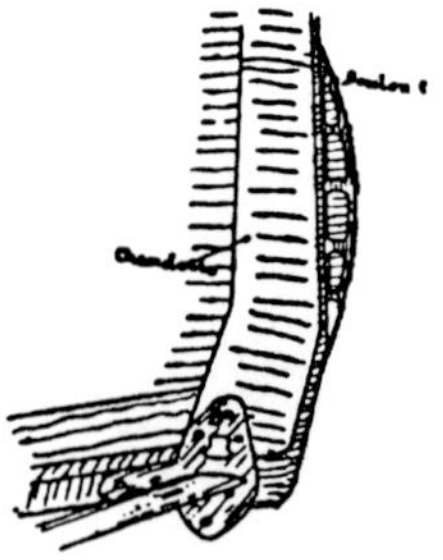

FORWARD ENGINE COMPARTMENT

SKETCHES FROM "MANUEL D'AVIATION de la PREMIER GUERRE MONDIALE: TOME II"

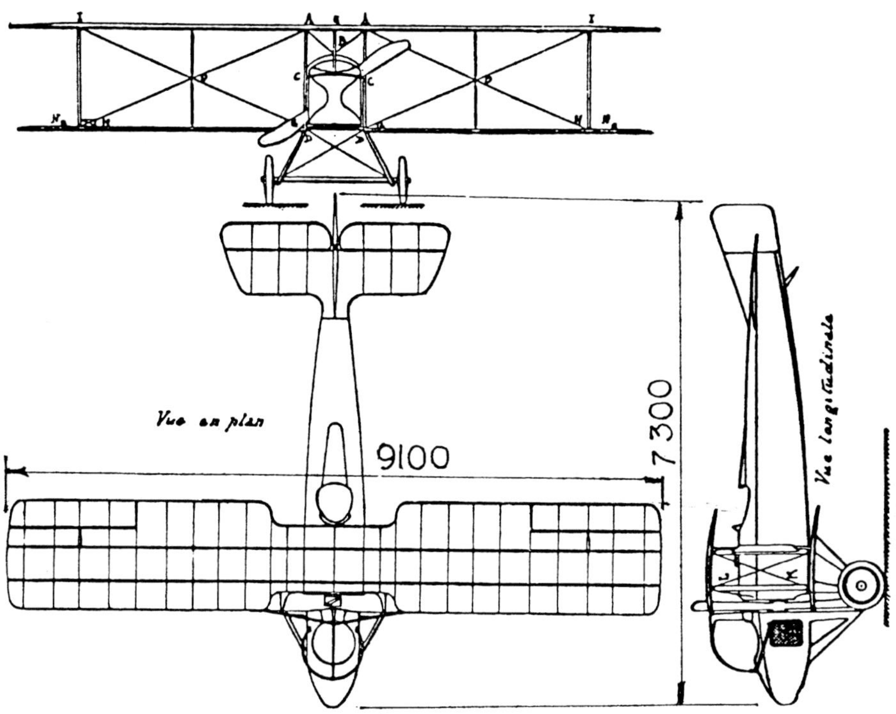

SA-I OR SA-2

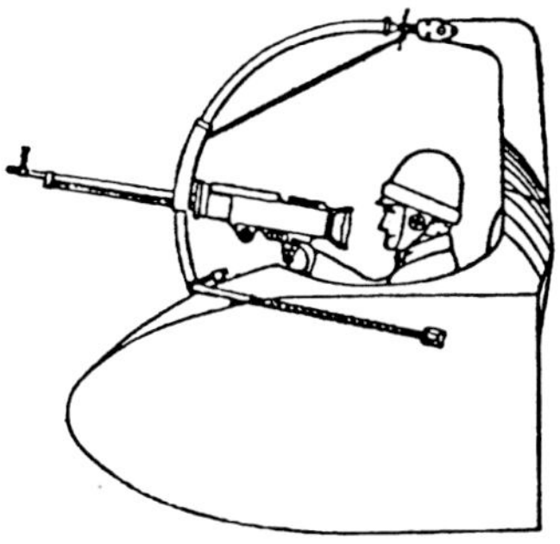

NACELLE

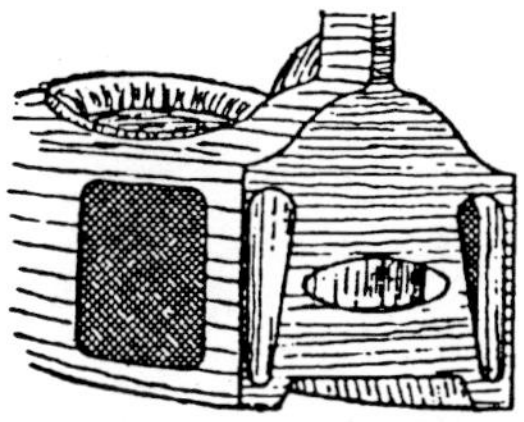

REAR OF NACELLE SHOWING COOLING HOLES

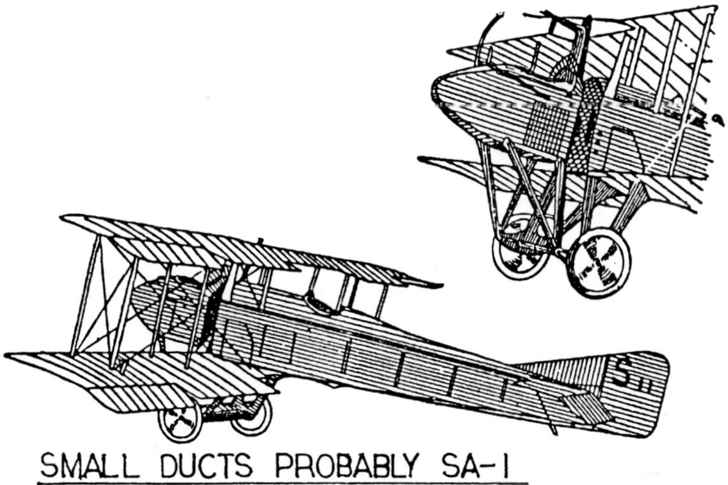

SMALL DUCTS PROBABLY SA-I

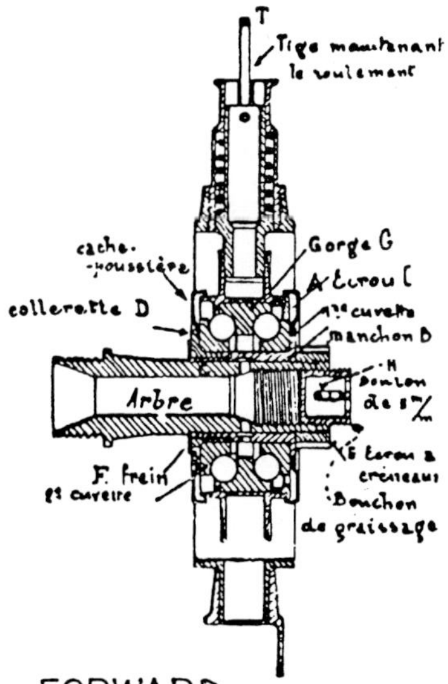

FORWARD →
NACELLE SUPPORT BEARING ASSEMBLY

Left: *The Französischer Kampffluzzeug SPAD* was captured nearly intact by the Germans in 1916. It had been in storage for years and had been a collection of parts. This was on display in the Deutsche Luftfahrt sammlung Berlin and is being viewed by visitors. (Gettyimages).

Below: *The Französischer Kampffluzzeug SPAD* was captured nearly intact by the Germans in 1916. It had been in storage for years and had been a collection of parts. The turtledeck and cockpit were incorrect, along with other errors such as the double bay bracing. This was on display in the Deutsche Luftfahrt sammlung Berlin (German Air Collection). (MVT)

Spad S.A.3 & Other Spad S.A.Derivatives

Above & Following Page, Above: S-40 was the only S.A.3. Based on the S.A.2 with 110 hp Le Rhône, it probably had manufacture's number 401. The pilot a mock-up Lewis machine gun and it had dual controls. (SHAA)

The S.A.3 was a single prototype with dual controls and a mock-up machine gun for the pilot. It had SFA number S-40 and the manufacturer's number was likely 401. Its airframe was a modified S.A.2 and it was powered by the 110 hp Le Rhône. The nacelle was broadened at the nose to accommodate rudder bars for the gunner and the cooling ducts were not as long horizontally but taller vertically than on the S.A.2.

The intention of the S.A.3 was to enable either crew member to act as gunner and the other as pilot depending on the position of the enemy relative to the aircraft. There appears to be a voice tube around the lower portion of the starboard under-carriage strut to help crew coordination as was tried on earlier S.A. aircraft.; its effectiveness is not known.

The S.A.3 was built because Béchereau hoped to quickly make the standard SA design more reliable. Development of the SA, originally planned to last 10 days, was not completed after a month. On 17 September a commission meeting at Chalais stated only that it would "at that point soon". Experience with the SA indicated to the design team and military authorities that the problems would never be solved.

The S.A.3 was also in response to an official G.Q.G. request for a fighter with more power and guns fore and aft made 2 August 1915. Although the authorities preferred the use of a twin-engine, three-place aircraft, the S.A.3 potentially could solve the desire with a smaller aircraft with a single engine.

The S.A.3 was doomed by the poor cooling of the engine and difficulty of the crew members coordinating in flight. It finished its life as a trainer.

Other designs Derived from the SA

There is no record of a rumored S.A.5.

The Spad SC. was a design study for a type derived from the S.A.The design had a 220 hp Renault V12 but was not built.

The 110 hp Clerget was considered as a replacement for the delayed 110 Le Rhône but was abandoned since the Clergets were reserved for Nieuport two-seaters from August 1915.

The Spad SD was a large, three-seat aircraft with fore and aft guns. Powered by a Panhard-Levassor of 240 hp, it was said to achieve 128 km/h with an endurance of 5

Above: The twin-engnie Ponnier P.1 with dual controls and two machine guns. Apparently it was designed to fulfill the same requirement as the S.A.3; only a single prototype was built.

Above: S-40 as a trainer; here at the far left of the lineup. (LGA)

hours. There are no confirmed photos of this aircraft but the aircraft on the facing page may be the SD.

From the beginning an alternative to the standard SA was an airplane that replaced the nacelle with gunner with a nacelle with up to four fixed guns fired by the pilot. The SA was tested with a mock-up nacelle and four fixed dummy guns. Encouraged to design a production aircraft, in the beginning of 1916 there were two formal designs, the SG-1 and SG-2. Formalized on 28 February 1916, the SG.1 were for two aircraft, retained the wingspan of the SA.2 but with reduced wing area since its gross weight was lower (660 kg versus 735 kg. The reduced wing area apparently was due to reduced chord on the lower wing.

The second version, the SG.2, was similar to the SG.1 but with reduced wingspan of 7.750 m compared with 9.550 m of the SG.1. Both designs were to be powered by a 110

Above: Close-up if the S.A.3 as a trainer.

Above: This large Spad featured a Spad nacelle; it was built by Blériot for the late 1916 bomber competition. The aircraft was a three-seater and the engine was a 240 hp Panhard. This aircraft may be the Spad SD. (SHAA)

Le Rhône. It is unknown whether either was built or flown. The performance figures are thought to be estimates, not measured performance.

SG.1 & SG.2 Project Specifications

Characteristic	S.G.1	S.G.2
Wingspan, Upper	9.555 m (31 ft. 4 inches)	7.550 m (25 ft. 9 inches)
Length	7.350 m (24 ft. 1 inches)	7.350 m (24 ft. 1 inches)
Height	2.600 m (8 ft. 6 inches)	2.550 m (8 ft. 4 inches)
Wing Area	24.10 m² (259.4 ft.²)	18.40 m² (198.1 ft.²)
Empty Wt.	440 kg (970 lb)	420 kg (926 lb)
Payload	220 kg (485 lb.)	
Total Wt	735 kg (1,620.4 lb.)	
Engine	110 hp Le Rhône 9J rotary	
Fuel	105 L (27.7 gal.)	
Oil	25 L (26.4 gal.)	
Gross Weight	660 kg (1,455 lb.)	640 kg (1,411 lb.)
Top Speed Estimated	161 km.h (100 mph) at 2000 m	
	154 km.h (95.7 mph) at 3000 m	
Climb Rate Estimated		
2000 m	7 min. 15 sec.	
3000 m	11 min. 30 sec.	

Spad S.A.4

Above: S-67 was a typical S.A.4. It was assigned to the 2nd KAO of the 1st BAG in summer/autumn 1916.

Russia had very limited aircraft and aeroengine production in WWI, and France was a major foreign supplier for both. Thus Russia expressed an interest in the SA series and ordered 50 machines from France on 6 March 1916 at a cost of Fr. 17.000 each, or a total of Fr 850.000. The order appears to specify six SA.2 aircraft with 110 Le Rhône 9J and 44 SA.4 aircraft with 80 hp Le Rhône 9C engines. Six additional SA.4 airframes without engines were ordered directly from the SPAD company. The 56 machines left the factory for Russian be the end of June 1916. The S.A.2 aircraft likely had SFA numbers S-42 through S-47 and the SA.4 aircraft likely had SFA numbers S-48 through S-97.

However opening of Russian records after the fall of the Soviet Union show slightly different number; 60 aircraft with 54 on a military order - 44 with the 80 hp engine (S-48 thought S-91) and 10 with the 110 hp engine. The aircraft with 110 hp engines may have been the 10 additional S.A.2 aircraft on a French purchase order of 17 May 1916, S-92 through S-101) and six airframes without engines ordered directly from SPAD. These could be SFA numbers S-102 through S-107, although no SFA numbers may have been assigned.

The S.A.4 had the following differences from the S.A.2:
1. A center section which made assembly much easier.

S.A.4 Specifications	
Characteristic	
Wingspan, Upper	9.55 m (31 ft. 4 inches)
Wingspan, Lower	9.25 m (30 ft. 4 inches)
Chord	1.544 m (5 ft.)
Length	7.5 m (24 ft. 1 inches)
Height	2.65 m (8 ft. 8 inches)
Wing Area	24.5 m² (263.7 ft.²)
Empty Wt.	414 kg (912.7 lb.)
Useful Load	260 kg (573.2 lb.)
Armament	15 kg (33.1 lb.)
Crew	150 kg (330.7 lb.)
Fuel & Oil	95 kg (209.4 lb.)
Total Wt	734 kg (1,485.9 lb.)
Engine	80 hp Le Rhône 9C rotary
Ceiling	4,300 m (14,108 ft.)
Endurance	3 hr.
Top Speed	140 km.h (87 mph)
Climb Rate	
2000 m	13 min.

Above: Close-up of the engine of SA.4 S-54 after capture.

2. The lower wing ailerons were eliminated.
3. The two tie rods between the cabane struts were eliminated.
4. The chord of the lower wing was reduced.
5. The stabilizer chord was increased to 1.25 m and the small fairing was eliminated.
6. Moving the airfoil aft on the spars so that the structure of the fuselage would not be disturbed moving the wings aft, which reduced the tail heaviness.

The 110 hp 9J was just coming into production when the Russian order was made. The specification of the 80 hp 9C engine likely resulted in earlier delivery, which was urgent. The enemy deployed second line aircraft on the Eastern Fron, making maximum performance not as critical as on the Western Front. In addition, cooling problems were always a concern with the SA, and were even more of an issues with the 110 hp engine. Apparently S.A.2 S-48 was modified into the SA.4 with the changes noted above.

Below: S.A.2 S-46 was sold to the IRAS.

Spad S.A.4 (& S.A.2) in Russia

Above: S.A.2 S-47 from the 4th Artillery aviation detachment with a group of pilots, autumn 1916. This aircraft was flown by military pilot Shtabs-Rotmistr Mikhail Shirkov from September 1916 to April 1917. (Viktor Kulikov)

The Eastern Front was 600 miles wide and the Imperial Russian Air Service (IRAS) did not have enough aircraft to support it in 1916. Arrival of the SA series permitted a partial reorganization of its squadrons into specialized units to oppose the German air units more effectively.

Starting with the 18th KAO, which received S.A.4 S-53 on 18 June, more than 10 KAO received the S.A.2 and S.A.4 in the summer of 1916. The 18th KAO was followed closely by the 8th, 14th, 19th, 30th KAO, 1st Guards KAO, 2nd Siberian KAO, 4th Artillery Aviation Detachment (Art. AO), and 7th Army Aviation Detachment (Arm KAO). For example, the 4th Art AO received S.A.2 S-47 on 2 September and 7th Art. AO received S.A.4s S-95 and S-100 on 20 August.

By late 1916 it was obvious the SA was being outclassed by enemy aircraft even though the enemy was using second-rate equipment. The SA could interrupt reconnaissance planes away from their planned missions but for most ot the cases the two-seaters could outrun the SA fighters. Despite this, four claims of aerial victories attributed to Spad A aircraft can be identified in Russian archives.

On September 6, 1916, a German plane was seen around the airfield of 19th KAO based at Loutsk (today's Ukraine). The commander of the unit, the top Russian ace Shtabs Rotmistr A.A. Kozakov, took off on a single seat Nieuport X to attack it. During the fight, the German plane failed to notice the approach of a Spad A.4 from 19th KAO piloted by Podporuchik G.V. Bashinskiy with Podporuchik Boris A. Guber as observer. The enemy plane tried to escape but was shot down by the Spad, and fell in Berezoluky (Ukraine, 26 km NE Loutsk). The enemy flyers, Ltn Müller and Ltn Berger were wounded and burned their plane before being taken prisoners. This was the first victory by the group and likely by a Spad A in Russia.

A second victory possibly won by a Spad A from 19th KAO occurred on November 25, 1916, when Podporuchik V.V. Karpov (pilot) and Kornet G.A. Bratolyubov (observer) began an air battle with an enemy aeroplane between Loutsk and Rojychtche (Ukraine). As they met, the German opened fire first and damaged the Russian plane's left wing with explosive cartridges. After Bratolyubov fired an entire clip the German lost altitude. After locating him, the Russian crew fired off another clip and a half. The enemy then descended steeply near the ground and began to dodge, landing near a Russian artillery brigade. The enemy crew escaped after burning their plane, an Aviatik with a Mercedes 160 hp engine, who was recovered by Russian troops with a wing burnt. The fact that the Russian crew was flying a Spad remains unsure, and the claim cannot be identified in German archives.

On March 3, 1917, Shtabs Kapitan Kotov (pilot and commander officer of 2nd Siberian KAO), flying with observer Kornet Dilinka, claimed while flying a Spad A.2 an enemy aircraft shot down near Dusche. The claim cannot be confirmed in German or Austro-Hungarian archives.

Then, on May 5, 1917, Spad A.4 S-79 from 30th KAO piloted by Efrietor E.S. Nedzevetskiy with Starshiy unterofitser Rodin as observer, while protecting a photo-reconnaissance flight, met an Austro-Hungarian Hansa-Brandenburg tow-seater and attacked it at Gura Humorului (Romania). The enemy plane was Brandenburg C.I (U) 63.23 of Flik 1. The pilot, Leutnant Vlastimil Fiala, and observer Oberleutnant Johann Baumgartner, were captured.

During the winter 1916-1917, IRAS started to receive Nieuports and later Spad VII in numbers. This allowed the elite units to transfer their SA machines to less prestigious units. The more elite units replaced their last SA aircraft in the spring of 1917.

On May 28, 1917, S.A.4 S-64 was shot down by three Fokker Eindeckers, killing the Russian crew, Feldwebel Bigel (pilot) and Podporuchik Tarkho (observer), from 27th KAO, crashing in Focsani (Romania). The fixed, synchronized guns of the Fokkers made the difference because they were not stellar performers. The Fokkers were also obsolete on the Western Front a year previously.

Last mention of a Spad A in Russian archives came on June 27, 1917, when S-54 landed by mistake behind the Austrian lines in Krasne, 30 km east of Lviv (Ukraine), and was extensively photographed. According to Austrian documents, pilot Jemeljan Kusmitsch Myskin and mechanic Valentin Konstantinovich Ledilin were taken prisoners. The plane was from 1st Turkestan KAO based at Loutsk (Ukraine).

Altogether, at least 17 Russian units used the Spad SAs, some to the end of hostilities. The SA soldiered on after the armistice for training. The Russians used the SA series on skis during the winter. Russian aviators had the same complaints as the French aviators. The nacelle support system was complex and could fall off if damaged in combat. One observer had his neck broken when his scarf became entangled in a rotating propeller.

Summary

Fokker Eindeckers with synchronized guns arrived at the front essentially simultaneously with the SA, making the SA obsolete on arrival. The SA was not a success on operations However, the basic structural design was very strong and with the proper modifications and addition of the Hispano-Suiza V-8 engine, SPAD developed the Spad VII, one of the most successful fighters of the war and sire of the Spad XIII, widely acknowledged as the Allies' best fighter.

Above: Military pilot Ensign Ivan Vinogradov is standing near his S.A.4, and Cornet Anatoly Vershkovsky is an observer in the cockpit. 30th KAO, summer of 1916. (SHAA)

Above: S.A.4 fitted with skiis. The omision of ailerons on the lower wings confirms it is an SA.4. (SHAA)

Above: S.A.2 in Russia.

Above: Senior non-commissioned officer pilot of the 19th KAO Vladimir Shaitanov in front of his Spad S.A.4. The observer in the front cockpit is armed with a Vickers machine gun. On the rudder is a unit marking – "Adam's head" (symbol of immortality) on a black background. The airfield near the city of Lutsk, autumn 1916.

Below: An S.A.4 in Galicia with members of an unknown unit.

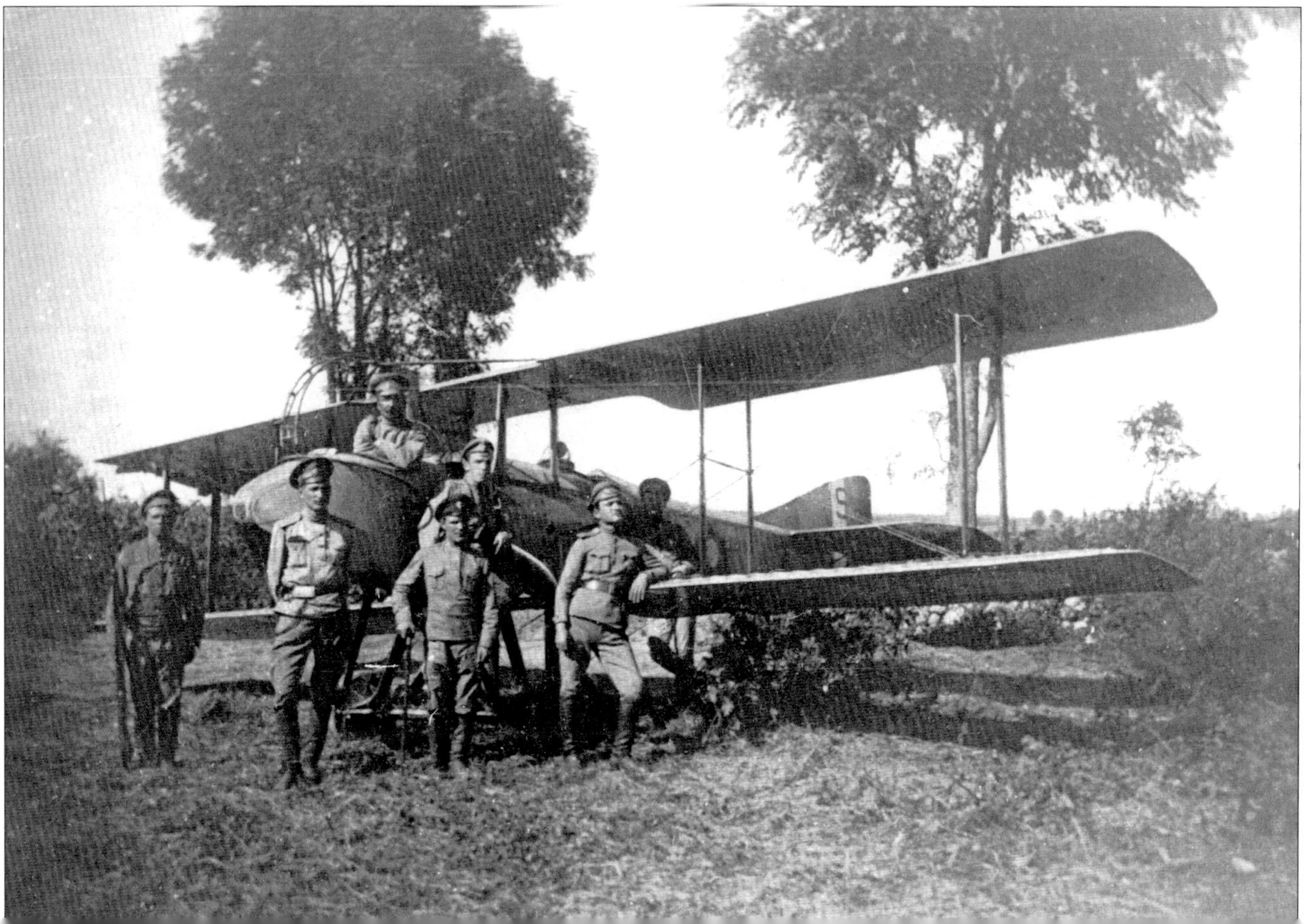

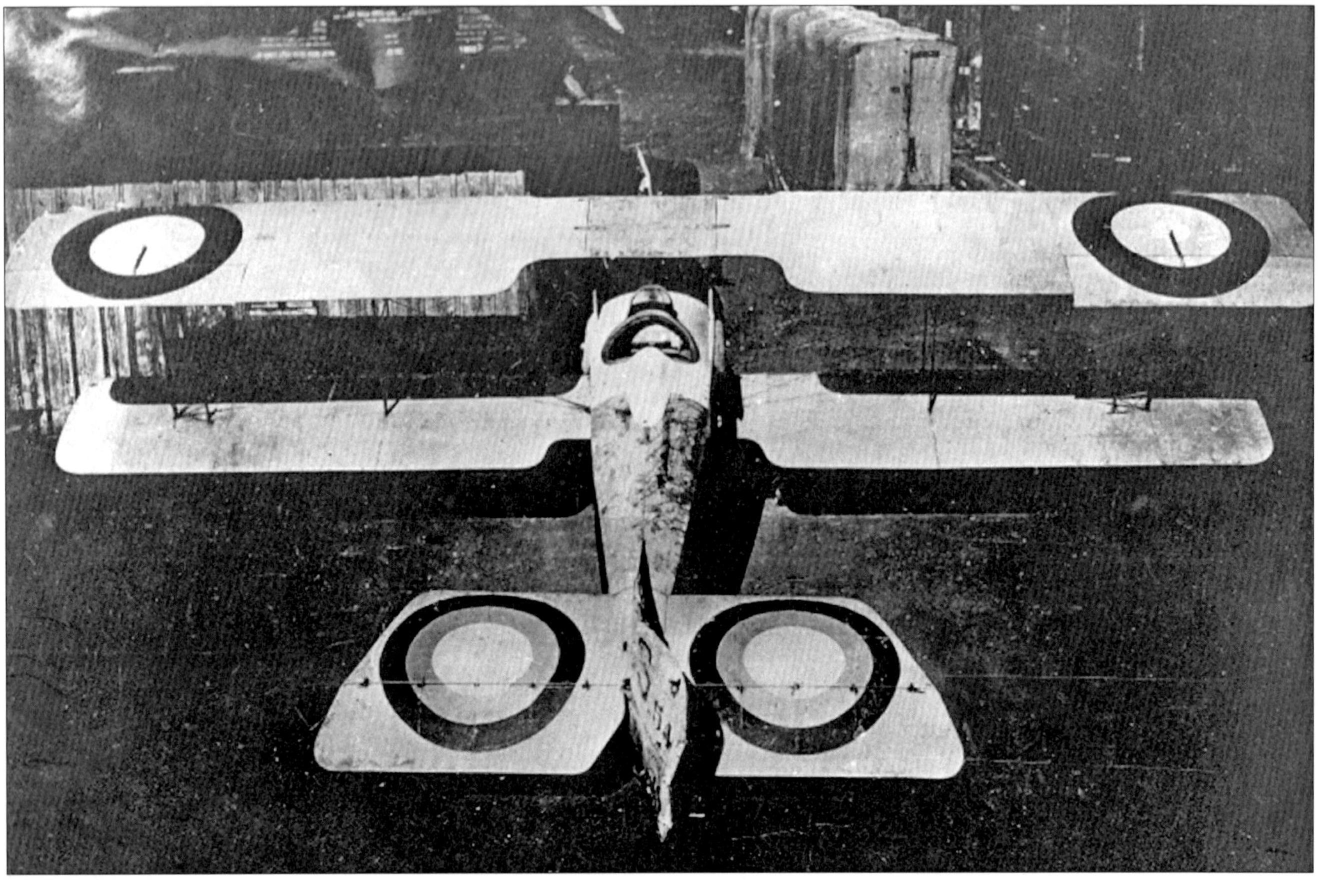

Above & Below: On 15 June 1917 S.A.4 S-54 landed in Austrian lines by mistake and was captured. It created a sensation in the Austrian troops and was widely photographed. The extensive use of national markigs is a omment on recognition standards of the time. (SHAA)

Above: S.A.4 S-54 being transported to Flik 14, then on to Aviation Part 4. (SHAA)

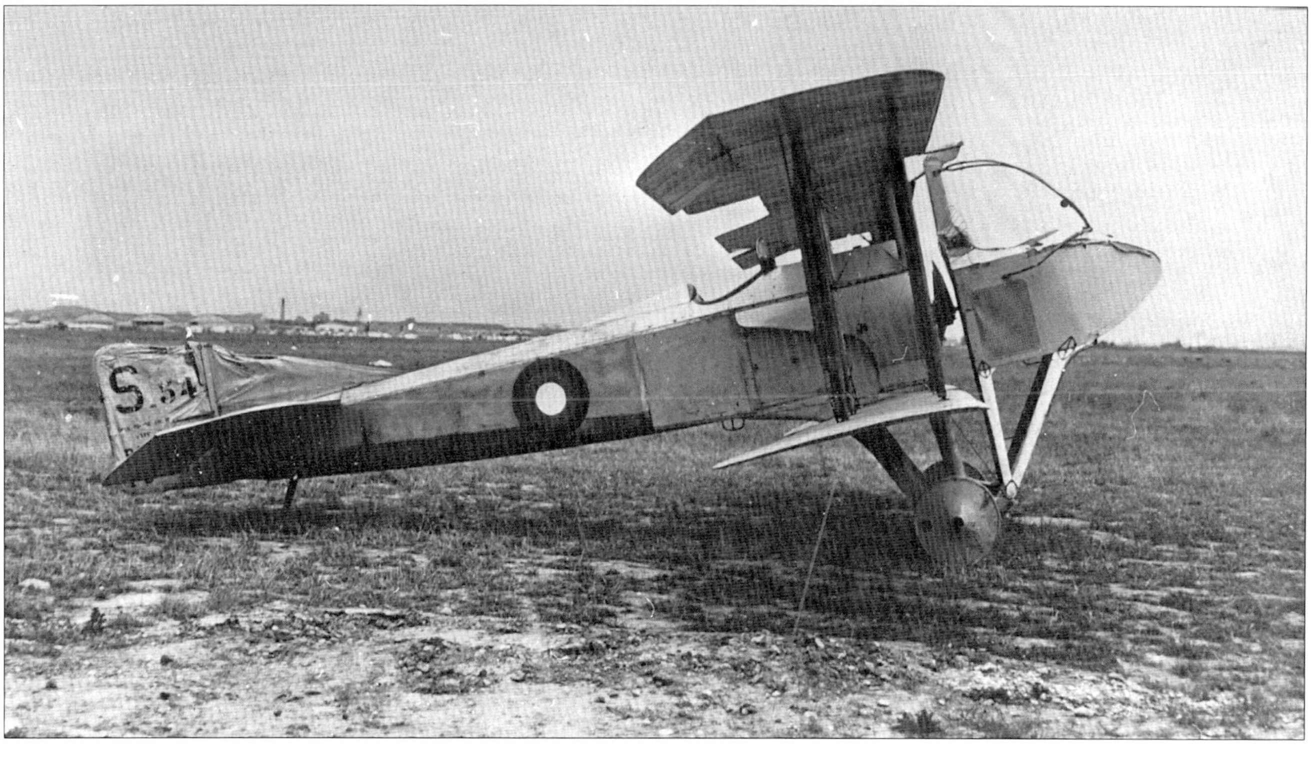

Above: Another view of S.A.4 S-54after capture. (SHAA)

Above & Below: S.A.4 S-54 captured after landing in Austrian lines. (SHAA)

Above: Another view of S.A.4 S-54 captured after landing in Austrian lines. (SHAA)

Above: S.A.4 S-60 from the 30th KAO. This aircraft entered the detachment on July 22, 1916 and was in service until the end of the year. It was flown by senior non-commissioned officers Ryabokon and Kibasov and Ensign Petrov. (Viktor Kulikov)

Above: S.A.4. S-67 of the 1st BAG at Loutsk, September 1916.

Above: S.A.4. S-67 from the 2nd KAO 1st BAG. Podporuchik Ivan Babkin, a military pilot, is sitting in the cockpit. The observer Poruchik Alexander Akaro has a Colt machine gun on a standard installation. The airfield near Lutsk, autumn 1916.

Above: S.A.4 S-71. Guards Corps aviation detachment in a canvas tent after assembly. November 1916. Senior non-commissioned officer Roman Yarmorkin flew it until June 1917. (Viktor Kulikov)

Above: S.A.4 S-75 from the 2nd KAO, the 1st BAG, autumn 1916. In February 1917, it was transferred to the 26th KAO. (Viktor Kulikov)

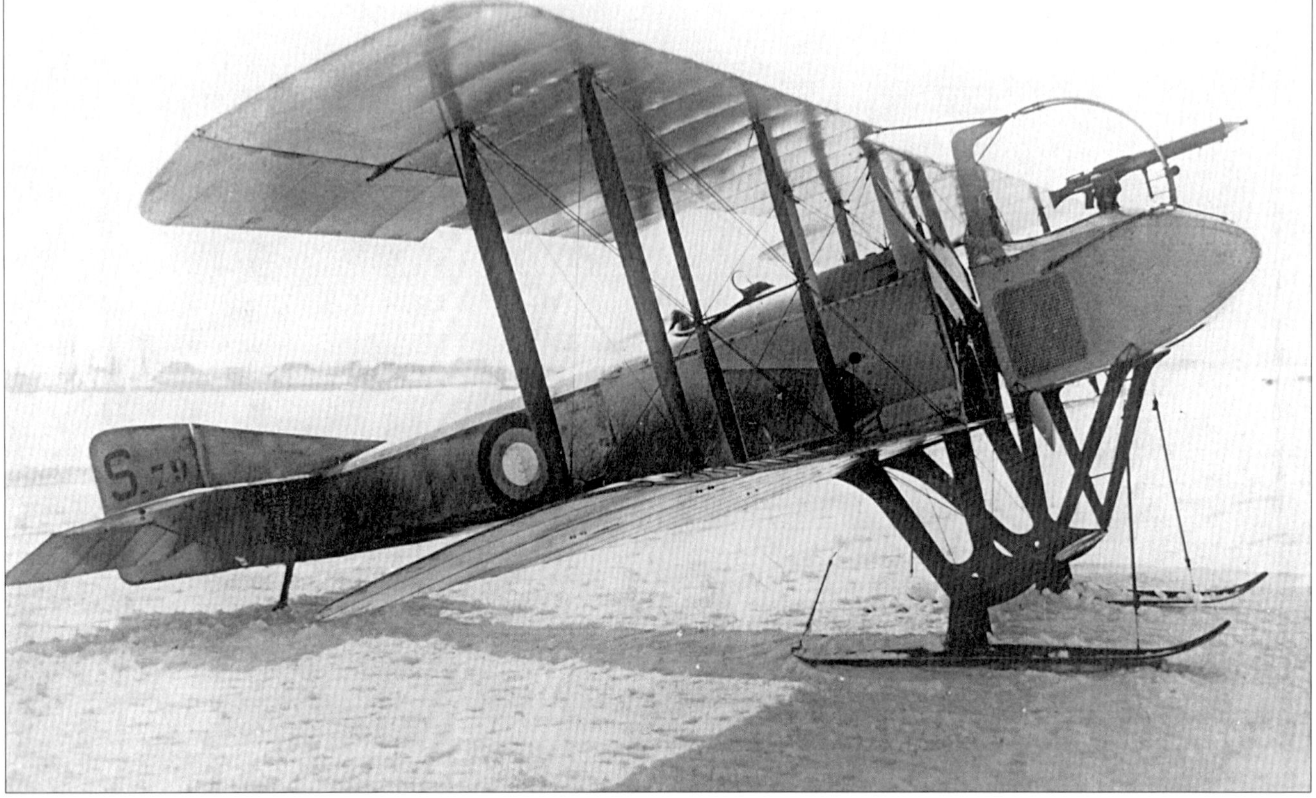

Above: S.A.4 S.79 on 04/22/5/17: the pilot, private soldier Evgeny Nedzevetsky and the observer, senior non-commissioned officer Prokofy Rodion, both from the 30th KAO, shot down a Brandenburg. Both Austrian officers were captured; the pilots managed to burn the plane. The enemy plane was shot down in the Gura-Gumora area. Brandenburg C.I N 63.23 pilot Lieutenant of reserve Vladastimil Fiala, observer Oberleutnant Johan Baumgartner, were both from the Austrian aviation company Flik1.

Above: A crewman poses with his S.A.4 S.79 fitted with skiis. (SHAA)

Above: S.A.4 S-79 form the 30th KAO. Summer 1916. (Viktor Kulikov)

Above: S.A.4 S-80 form the 30th KAO. Military pilot Ensign Nikolai Petrov fought in it from July 1916 to April 1917. (Viktor Kulikov)

Military pilots of the 30th KAO and an S.A.4 on skis: Warrant officers Ivan Vinogradov, unknown, Podporuchik Alexander Shishkovsky. January-February 1917. (Viktor Kulikov)

Above: Military pilot of the 30th KAO, Podporuchik Ivan Vinogradov in the cockpit of an S.A.4. Autumn of 1916. (Viktor Kulikov)

Above: S.A.4 before departure for a combat mission. There is a Lewis infantry machine gun in the observer's cabin. (Viktor Kulikov)

Above: Observer Ensign Georgy Bratolyubov in an S.A.4 from the 19th KAO, November 1916, Lutsk. Note the unusual modification of the Lewis infantry machine gun in the gunners' cockpit. (Viktor Kulikov)

Above: The breakdown of S.A.4 from the 2nd KAO is the 1st BAG. Autumn of 1916. (Viktor Kulikov)

Above: The adjutant of the 14th KAO, Poruchik Vasily Bekarevich, with the sisters of mercy – a souvenir photo of the S.A.4. The white rudder is without the usual markings. (Viktor Kulikov)

Above: A crewman poses with his S.A.4. (SHAA)

This Page: Three photos of S.A.4 S-87 being readied for flight. The gun is an infantry mode Lewis. The front plate had no extra lightening holes (SHAA)

Above: S.A.4 S-87, a typical Russian S.A.4. (SHAA)

Above & Below: Military pilot Vladimir Shaitanov and observer Ensign Georgy Bratolyubov in front of their S.A.4 (above) and in their aircraft (below) from the 19th KAO. November 1916, Lutsk. (SHAA)

Above: S.A.4, probably S.87, with the observer's cockpit tilted forward after a hard landing, autumn 1916. This aircraft was flown by pilot Corporal Vladimir Anders from October 1916 to June 1917. (SHAA)

Right: S.A.4 in Russia with animal marking on the side with nacelle down for maintenance.

Above: Lineup of the 1st BAG at Loutsk in September 1916. Spad S.A.4s are intersperced with Nieuport 10s.

Above: Pilot Senior non-commissioned officer Grigory Kibasov (wearing a flight helmet) near the wrecked S.A.4 S-84, which was destroyed on May 20, 1917. Forced landing due to poor engine operation outside the airport and nose over. Observer Abramov was seriously injured.(Viktor Kulikov)

Above: Crash of S.A.4 S-82 in Russia. (A. Toelle)

Above: Fatal crash of a S.A.4, positvely indentified by the lack of an aileron on the lower wing. The bodies of the crew are visible. This may be the crash of S-64 of the 27th KAO shot down in Romania on 16/29 May, 1917, by three Fokker Eindeckers. Both types had disappeared from the Western Front a year earlier. (SHAA)

Above: A typical Russian S.A.4 mounted on skiis. (SHAA)

Below: S.A.4 S-90 in IRAS service somewhere on the front. (SHAA)

Above: IRAS squadron lineup with SA.4s and, on the far right, two rare Vickers FB-19s, six of which were purchased by Russia, (SHAA)

Above: IRAS squadron lineup with S.A.4s. (SHAA)

Right: Fatal crash of a Russian S.A. (SHAA)

Left: S-97 was an S.A.4 that was modified into a single-seat fighter by replacing the gunner and his flexible gun with three fixed Colt machine guns. S-97 is the highest SFA number for which we have a photograph that can be identified and is one of the ten S.A.4 fighters that had the110 hp Le Rhône 9J engine. The result was very similar to Béchereau's original idea of four fixed machine guns fired by the pilot in a modified nacelle without gunner. This aircraft was flown by a military pilot of the Guards Aviation Detachment, Podporuchik Ivan Kortaci, from October 1916 to May 1917.

Below: Side view **of** S-97 after the S.A.4 was modified.

Above: Front view of S-97 showing the three fixed Colt machine guns.

Right: Side view **of** S-97 after the S.A.4 was modified showing the Bowden cable that fired the guns going up the aft nacelle support. (E. Sommer).

Above: S.A.4 in IRAS service.

Above: S.A.4 in Soviet service.

Above & Below: An S.A. being modified in Russia (above) and after completion (below into a single-seat fighter). This paralleled the similar design changes the SPAD company made to develop the prototype of the excellent Spad VII with the exception that the SPAD company had access to the more powerful Hispano-Suiza V8 engine. The Spad A series was not an operational success but its structure formed the basis of the excellent Spad VII, which in turn led to the Spad XIII, widely considered the best Allied fighter and arguably the best fighter of the war.

Notes on Spad S.A.1 Series Aircraft

SFA No.	Spad No.	Type	Remarks (See Note 15 for *1, *2, *3, etc.)
S-1	401	S.A.1	First example of the series. It had a nonstandard nacelle. It was sent to Escadrille N 23 at Chalons-sur-Marne in August 1915 and flown by Capitaine Maurice Schlumberger. *1, *2
S-2		S.A.1	
S-3		S.A.1 (13)	It was at the Ecole de Pau in August 1916, and a photograph shows very well used fuselage fabric. *1, *2
S-4		S.A.1 (13)	It was in the Réserve Générale de l'Aéronautique in October 1915 and then modified with the round ducts, double-tube machine gun mount, and the later rudder inscription. *1, *3
S-5		S.A.1 (13)	In an unknown escadrille in the autumn of 1915.
S-6		S.A.1 (13)	In Escadrille N 23 in November 1915; flown by Sergent Jacques Sénart.
S-7		S.A.1 (13)	At Buc in September 1915 with an artificial, wooden Lewis gun, and then modified with the round ducts and an engine mount plate with large cutouts for better cooling, but it retained the single-tube machine gun mount. *1,
S-8		S.A.1	
S-9		S.A.1 (13)	Attached to Escadrille N 3 at Breuil-le-Sec by the end of 1915 and flown by Capitaine Félix Brocard. *1 and undoubtedly *2, but at some point the rudder inscription was changed to the later form P.U. ?: P.C.95. It was flown by Sergent Antoine Richard with a cocarde on the nacelle and a small tricolor flag painted on the top of the rudder. Sergeant Richard and his gunner, Soldat Pillon, were killed in an accident on 8 March 1916.
S-10		S.A.1 (13)	Attached to Escadrille N 23 during the autumn of 1915. There seems to be vertical, tricolor stripes on the rudder, a cocarde on the fuselage, and a white lozenge with a red border on the nacelle. Crashed behind German lines on 22 November 1915 at the Aure after an aerial combat. Both Capitaine Schlumberger, pilot, and Capitaine Montezuma, gunner, were killed.
S-11		S.A.1 (13)	First flown by Maurice Prévost on 21 May 1915. Apparently retained by the SPAD company and flown with a modified S.A.1 nacelle and cowl but still with the original triangular-shaped tailplane. It was sketched as the S.A.1 standard for the *Manuel de Montage du Spad A* dated 1916. The sketch shows the trapezoidal trailpland and *1. *2

Notes on Spad S.A.2 and 3 Series Aircraft

SFA No.	Spad No.	Type	Remarks (See Note 15 for *1, *2, *3, etc.)
S-12		S.A.2 (13)	A photograph shows tricolor chevrons on the fuselage. Possibly assigned to C.R.P. *5
S-13		S.A.2 (13)	Aircraft in school in 1916, ID number 12. Tricolor chevrons on fuselage, ex C.R.P.(?) *4, *6
S-14		S.A.2	
S-15		S.A.2	
S-16		S.A2 (13)	Only P.C. 100 can be seen on the rudder.
S-17		S.A.2 (13)	Only P.C. 100 can be seen on the rudder and there were vertical tricolor bands on the fuselage and a Morane type L machine-gun mount on the nacelle.
S-18		S.A.2	
S-19		S.A.2 (13)	Inscribed *MA JEANNE* on the fuselage, with the ID number S-19/10-15 II burned into the longeron. *4, *6
S-20		S.A.2	
S-21		S.A.2	
S-22		S.A.2	
S-23		S.A.2	
S-24		S.A.2 (13)	Possibly sent to Escadrille N 3 at Cachy in May 1916 and overturned by Lieutenant Albert Deullin.
S-25		S.A.2	
S-26		S.A.2 (13)	*4, *6
S-27			
S-28			
S-29			

S-30			Assigned to Escadrille N 37 and flown by Sous-Lieutenant Pierre de Jumilhac from 8-24 April 1916, at which date it was removed from the front due to wear and tear.
S-31			Assigned to Escadrille N 73 and flown by Caporal Robert Sabatier with gunner, Caporal Théophile Funck-Brentano. They secured a victory over an Aviatik C on 17 May 1916 at Chatas. It was the last French type S.A. on the Western Front and probably stayed with N 73 until 1 July 1916. *4, *6
S-32			
S-33		S.A.2 (13)	No photographs of the rudder available, so the identification comes from the SFA number burned into the nacelle longeron, S-33/11-15 IL. It had a cocarde on the nacelle, and there was an extra bracing cable from the landing gear to the middle of the lower wings, and the front engine mount was cut out for better cooling. *4, *6
S-34			
S-35			
S-36			
S-37			
S-38			
S-39		S.A.2 (13)	Tricolor flag painted on the fuselage. *4, *6
S-40		**S.A.3 (13)**	Prototype of the dual-control machine with a machine gun for the pilot as well as one for the gunner. It was first flown on 21 February 1916 and was in school very late in 1916. *4, *6
S-41		(S.A.2) ?	Possibly modified to be the prototype for the SG-1, which was a single-place fighter with a nacelle. Existing drawings show a machine with a bullet-shaped nacelle and a 110hp Le Rhone rotary engine. The drawings show two possibilities differing in wingspan.
S-42	446	S.A.2	First S.A. type sent to Russian 14th KAO on 17 December 1916, but the engine did not arrive until 1 February 1917.
S-43	447	S.A.2	
S-44	448	S.A.2	Sent from 4th KAO to 1st BAG on 1 November 1916, then to 14th KAO on 29 May 1917, and flown by Sergeant Gude. It had flown 55 minutes by 1 June 1917.
S-45	449	S.A.2	
S-46	450	S.A.2 (13)	Without a doubt , equipped with an 80 hp Le Rhône in lieu of the 110 hp Le Rhône.
S-47	451	S.A.2 (13)	Moved to 4th Art. KAO on 2 Sept. 1916; flown by Staff Capt. Shirkov. Had flown 24 hour 20 minutes by 1 April 1917.

Notes on Spad S.A.4 Series Aircraft

SFA No.	Spad No.	Type	Remarks (See Note 15 for *1, *2, *3, etc.)
S-48	Unk.	S.A.4	Probably an ex-S.A.2 taken from the end of the line to test the modifications required to produce the S.A.4. Afterward it received the livery of the IRAS.
S-49	452	S.A.4	
S-50	453	S.A.4	
S-51	454	S.A.4	At 2nd KAO of Ist BAG on 1 November 1916. Sent to 28th KAO on 21 January 1917 and flown by Lieutent Tikhomirov and Second Lieutenant Gurkov. It had flown 7 hours 2 minutes. by 1 April 1917.
S-52	455	S.A.4	At 4th KAO of Ist BAG on 1 November 1916. Sent to 27th BAG on 15 January 1917. Flown by Lieutenant Shishkovsky. It had flown 1 hour 16 minutes at 27th KAO.
S-53	456	S.A.4	Sent to 18th KAO on 23 June 1916. Flown by Ensign Konyukov and had flown 64 hours 10 minutes by 1 June 1917.
S-54	457	S.A.4 (13)	At Turkestanskii KAO. It was lost and landed within the Austrian lines near Krasne, and the aircrew was taken prisoner on 15 June 1917. (Russian sources mention 18 June but that is probably the date of the report of loss.) It was examined by *Flik* 14, then sent to Aviation Park 4. Alan Durkota interviewed Rudolph Simacek, who was present at *Flik* 14 when the aircraft was brought in, and Simacek recalled that it created a lot of excitement and people took many pictures, which now surface all over the world.

S-55	458	S.A.4	At 4th KAO on 1 November 1916.
S-56	459	S.A.4	Sent to 30th KAO on 8 September 1916. Flown by Ensign Zalozny. It flew 48 hours by 1 March 1917.
S-57	460	S.A.4	Sent to 30th KAO on 6 September 1916 and flown by Lieutenant Vinogradov and Second Lieutenant Zaloz had flown 72 hours 50 minutes by 1 March 1917.
S-58	461	S.A.4	Sent to 8th KAO on 7 July 1916 and flown by Ensigns Ingistov and Egor'ev. It had flown 49 hours and 49 min by 1 April 1917. Sent for repair in May 1917.
S-59	462	S.A.4 (13)	Sent to 18th KAO on 6 September 1916 and flown by Corporal Vishnevsky. It had flown 12 hours 24 minutes by 1 November 1916 and continued in service until 1 March 1917.
S-60	463	S.A.4	Sent to 30th KAO on 22 July 1916; flown by Ensign Petrov and Sergeant Kibasov. It had flown 15 hours by 1 July.
S-61	464	S.A.4	Sent to 30th KAO on 22 July 1916 and flown by Ensign Vinogradov. Had flown 1 hour 30 minutes by 1 November 1916.
S-62	465	S.A.4	Sent to 8th KAO on 7 July 1916 and flown by Second Lieutenant Rudlizky and Ensign Ingistov. It had flown 85 hours 40 minutes by 1 April 1917. It was sent to 1 Turkestanskii KAO in April 1917. It was flown by Sergeant Kondakov and was sent to repair in May 1917.
S-63	466	S.A.4	It was sent to 30th KAO on 22 July 1916 and flown by Ensigns Alelyukhin and Vinogradov and Sergeant Kibasov. It had flown 73 hours 50 minutes by sometime in 1917.
S-64	467	S.A.4	At 19th KAO of Ist BAG on 1 November 1916. Sent to 27th KAO on 31 January 1917. It was flown by Ensign Z. and Lieutenant Shishkovsky. It had flown 7 hours 10 minutes by 1 April 1917. It was shot down on 16 May combat with 3 Fokker E types in the region of Bileshti in Romania. Both the pilot, Sergeant Bigel, and gunner, Lieutenant Tarkhov, were killed.
S-65	468	SA.4	
S-66	469	S.A.4	Sent to 28th KAO on 15 March 1917. Flown by Second Lieutenant Gurkov for 4 hours 50 minutes by 1 June 1917.
S-67	470	S.A.4 (13)	Partial rudder inscription P.C. 95. At 2 KAO of 1 BAG in August 1916 with Colt machine gun. At 26th KAO on 1 March 1917.
S-68	471	S.A.4	At 4th KAO of Ist BAG on 1 November 1916, at 14th KAO on 5 January 1917. Flown by Sergeant Gude.
S-69	472	S.A.4	Sent to 30th KAO on 22 Jul 1916. Flown by Sergeant Khovostov. It had flown 95 hours 10 minutes by 1 April 1917.
S-70	473	S.A.4	Sent to 4th KAO of Ist BAG on 29 July 1916. Sent to 30th KAO on 31 May 1917. Flown by Staff Captain Studzinsky and Ensign Vinogradov. It had flown 93 hours 20 minutes by 1 June 1917.
S-71	474	S.A.4	Sent to Ist Guard KAO in October 1916. Flown by Sergeant Yarmorkin. It had flown 7 hours 50 minutes by 1April 1917. Sent for repair on 1 June 1917.
S-72	475	S.A.4	At 18th KAO on 1 November 1917.
S-73	476	S.A.4	At 4th KAO of Ist BAG on 1 November 1916.
S-74	477	S.A.4	At 2nd KAO of Ist BAG on 1 November 1916. Sent to 26th KAO on 10 January 1917. Flown by Captain Antono and Ensign Petrozhitsky. It had flown 2 hours and 7 minutes by 1 March 1917.
S-75	478	S.A.4	At 2nd KAO of Ist BAG in August 1916. Sent to 28th KAO on 6 February 1917. Flown by Second Lieutenant Petrov. It had flown 17 hours 45 minutes by 1 June 1917.
S-76	479	S.A.4	At 30th KAO on 1 November 1916. Sent to 28th KAO 6 December 1916. Flown by Sergeant Tsvetko and Second Lieutenant Tikhomirov. It had flown 20 hours 35 minutes by 1 June 1917. It was still in use in September 1917.

S-77	480	S.A.4	Sent to 8th KAO on 17 July 1916. Flown by Corporal Malyarevsky; it had flown 73 hours 45 minutes by 1 March 1917. It had no engine by 1 April 1917.
S-78	481	S.A.4 (13)	Sent to 8th KAO on 17 July 1916. It was flown by Staff Captain Kotov and Second Lieutenant Rudlitsky. It Turkestanskii KAO and flown by Corporal Tsedilin on 1 May 1917. It had 54 hours 10 minutes by 1 June 1917.
S-79	482	S.A.4 (13)	Sent to 30th KAO on 22 July 1916 and flown by Staff Captain Studzinsky, Ensign Vinogradov, and Private Nedzevetsky. It had flown 29 hours 45 minutes by 1 November 1916. It was on skis with a Lewis machine gun during the winter 1916-1917. One victory secured over Brandenburg C.I (U) 63.23 of *Flik* 1. The pilot, Leutnant Vlastimil Fiala, and observer Oberleutnant Johann Baumgartner, were captured on 22 April/15 May 1917. Pilot, Private Nedzevetsky, and gunner, Corporal Rodin, got the credit. The aircraft went in for repair in June 1917.
S-80	483	S.A.4	Sent to 30th KAO on 22 July 1916 and flown by Staff Captain Studzinsky, Ensign Vinogradov, Ensign Petrov. It had flown 138 hours 55 minutes by 1 April 1917.
S-81	484	S.A.4	Sent to 18th KAO on 31 July 1916 and flown by Staff Captain Studzinsky and Second Lieutenants Toropov and Ignatiev. It had flown 54 hours 45 minutes by 1 June 1917 and then went for repair at 4 Aviation Park on 1 July 1917.
S-82	485	S.A.4	At 4th KAO of Ist BAG on 1 November 1916 and then sent to 27th KAO on 15 January 1917 and flown by Sergeant Bigel. It flew 16 hours 48 minutes at 27 KAO, and then it was sent to lst Air Park for repair on 1 May 1917
S-83	486	S.A.4	Sent to 30th KAO on 30 June 1917 and flown by Ensign Zalozny. Flown 4 hours 25 minutes by by 1 July 1917.
S-84	487	S.A.4	Sent to 30th KAO on 18 December 1916 and flown by Ensign Vinogradov. It had flown 37 hours 50 minutes by 1 April 1917 and went to 6th Air Park on 1 June 1917.
S-85	488	S.A.4	On charge with Ist Aviation Division on 1 August 1917. It was burned during the retreat from Tarnopol in July 1917.
S-86	489	S.A.4	Sent to 14th KAO on 5 January 1917. Flown by Ensign Alelyuhin. It had flown 20 hours 3 minutes by I June 1917.
S-87	490	S.A.4 (13)	Sent to 18th KAO on 7 October 1916. Equipped with Lewis machine gun. Flown by Private First Class Anders. It had flown 63 hours 45 minutes by 1 June 1917. *1, partial rudder inscription P.C. 95.
S-88	491	S.A.4	Sent to 30th KAO on 18 December 1916 and flown by Sergeants Ryabokom and Khvostov. It had flown 63 hours 25 minutes by 1 June 1917.
S-89	492	S.A.4	At 2nd KAO of Ist BAG on November 1916. Sent to 14th KAO on 5 January 1917. Flown by Ensign Alelyukim. It had flown 10 hours 28 minutes by | April 1917 and then sent to 30 KAO on 28 August 1917 and flown by Ensigns Petrov and Zalozny.
S-90	493	S.A.4	At 4th KAO of Ist BAG on 1 November 1916 and then sent to 14th KAO on 5 January 1917 and flown by Second Lieutenant Ivanov and Ensign Mashurko. It had flown 8 hours 6 minutes by 1 April 1917.
S-91	494	S.A.4	Sent to 14th KAO on 11 September 1916. Flown by Ensign Karklin and many other pilots.
S-92	495	S.A.4	Sent to 2nd Siberian KAO on 13 September 1916 and flown by Calvary Staff Captain Shirkov. It was at 30th KAO on 30 September 1917.
S-93	496	S.A.4	At 19th KAO on 1 November 1916.
S-94	497	S.A.4	Sent to 30th KAO on 15 January 1917 and flown by Ensigns Zalozny and Khvostov. It had flown 76 hours 30 minutes by 1 June 1917; still in use in September 1917.
S-95	498	S.A.4	Sent to 7th ArmAO (2nd Guards KAO) on 20 August 1916 and flown by Sergeant Kolev, Staff Captain Sakhnovsky, and Second Lieutenant Dorokhotov. It had flown 32 hours 23 minutes by 1 June 1917.
S-96	499	S.A.4 (13)	Sent to 1st Guards KAO in July 1916 and flown by Lieutenant Khrisoskoleo and Ensign Remmert. Last assigned to 6th ArmAO on 1 July 1917. *4(?), *6

1. All S.A.type Spads were constructed by the SPAD company.
2. All airplanes have two identification numbers: the constructor's and the service number supplied by the government. In France this number was supplied by the SFA (Service des Fabrications Aéronautiques).
3. The SFA number was applied to all major components. We are not sure where they were applied on S.A. aircraft, but based on later SPADs, these numbers would probably have been burned or stamped on the leading edge of the fin, the upper tip bow of the rudder, the right tip bows of the elevators and tailplane, and the underside of the tip of the forward spar of the right wing. We do know for sure that the SFA number was burned into the outside of the upper right longeron in the cockpit area of both the fuselage and the nacelle, along with the date of manufacture and a roman numeral denoting type. The number was covered with transparent plastic, and the fabric was cut away so that the numbers were visible. These numbers can be seen in photographs; although usually illegible because of the graininess of the photograph, we can read a few, such as S-33/11-15 II.
4. The first SFA number for the S.A. type aircraft was S-1. This number was usually painted on the rudder along with load information, although it was sometimes painted over.
5. It is not possible to match constructor numbers and SFA numbers before S-42, but S-1 through S-41 and S-48 (S.A.1, S.A.2, the prototype S.A.3, and the prototype S.A.4) plus one SPAD D (built before September 1915), and two SPAD E (construction started in November 1915) equal 45 aircraft. Constructor number 446 minus 45 gives the constructor number 401 for S-1. It should be noted that the SPAD company built Caudron aircraft under license at the beginning of the war, which explains why the S.A.series starts with 401.
6. The constructor numbers for S-49 through S-99 were probably 452 through 502.
7. The constructor number 503 was without a doubt assigned to the SPAD J, which was ready in April 1916.
8. The constructor numbers for S-100 through S-107 were probably 504 through 511.
9. The design changes for the new S.A.4 were undoubtedly tested on an aircraft taken from the end of the S.A.2 production line; therefore, S-48 had to have a constructor number lower than 446.
10. The prototype for the SPAD G-1 may have been S-41.
11. The constructor numbers for S-42 through S-47 were 446 through 451.
12. The SFA number for the prototype S.A.3 was S-40.
13. The aircraft is positively identified from photographs or documents of the period.
14. All the dates concerning the SPAD A aircraft in the IRAS are taken from Russian archives and are given in the Julian calendar dates of the period, which are 13 days behind the dates of the Gregorian calendar used by the West (e.g., the date 22 April 1917 Julian corresponds to 5 May 1917 Gregorian).
15. All S.A.1 aircraft were equipped with 80hp Le Rhone rotary engines, S.A.2 aircraft were equipped initially with 110hp Le Rhone rotary engines, and S.A.4 aircraft were equipped initially with 80hp Le Rhone rotary engines. Changes were occasionally made in service, and if we can document the change either with documents or photographs of the period they will be noted: *1 = 80hp Le Rhone, *4 = 110hp Le Rhone. Rudder inscriptions also indicate the engine installed: *2 = Charge Maxima 260, which usually indicates an 80hp Le Rhone; *3 = PU. 165 P.C.95, which also indicates an 80hp Le Rhone; *5 = Charge Maxima 280, which indicates a 110 hp Le Rhone; *6 = P.C.180 P.U.100, which also indicates a 110hp Le Rhone; and *7 = P.U.160 P.C.100, which indicates an 80 hp Le Rhone.

Spad SA Series Summary

The Spad A series was initiated at an early stage in the evolution of aviation technolgy, and suffered from it. In particular, the lack of a synchronizer for firing a machine gun safely through the propeller arc drove the Spad A configuration as a 'pulpit' fighter. Their technical success was limited by the arrival of Fokker fighters with sychronized guns virtually simultaneously with arrival with the SA at the front.

Additionally, the additional weight and drag of the necessary pulpit and second crewman seriously compromised the SA's performance. The gunner's nacelle also reduced the propeller effeciency and engine cooling, limiting the power of the engine that could be effectively used. These effects limited the Spad A's speed such that soon the aircraft it was attempting to intercept were faster.

As a result, the Spad A types are credited with a single confirmed victory and the Western Front (French service) and 2–3 confirmed victories on the less challenging Eastern Front (in Russian service).

In contrast to its mediocre operational contribution, its robust structural design formed the basis of all other SPAD designs that saw wartime serice. When a synchronizer was available to the SPAD company, the follow-on designs dispensed with the gunner's nacelle and replaced the gunner with a synchronized gun. Intallation of the more powerful Hispano-Suiza V8 instead of the rotary engine resulted in the Spad VII fighter, one of the war's most successful fighters. In turn it was developed into the Spad XIII fighter, the best French fighter that was a major factor in the Allied success in the air as it was supplied to the other Allied powers.

The SPAD SA's configuration was its problem and its structure was its contribution. The SA's arrival at the front simultaneously with enemies with synchronized guns was just unfortunate timing during a rapidly developing technology.

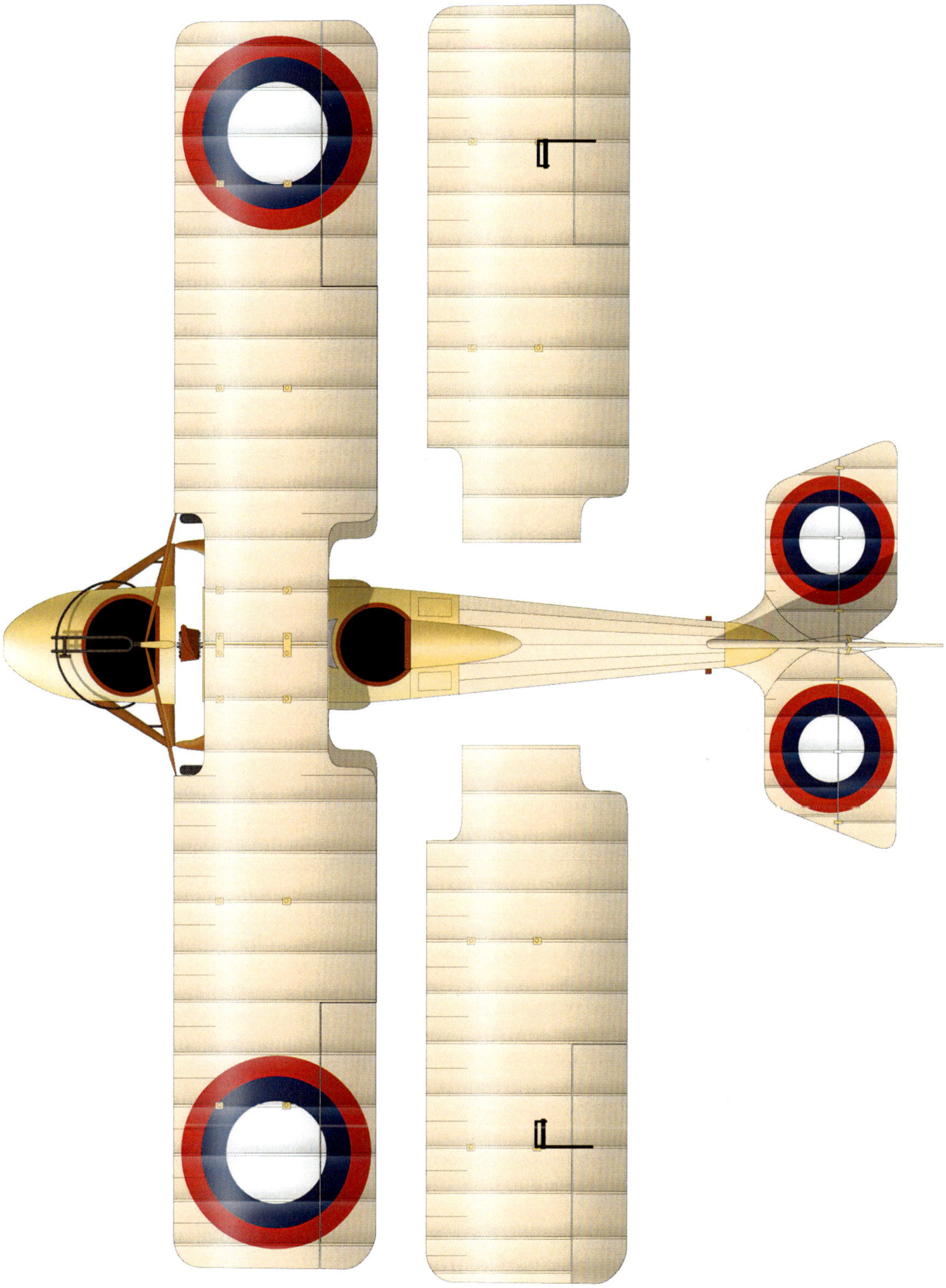

Spad S.A.2 IRAS Plan View
Spad S.A.4 IRAS similar

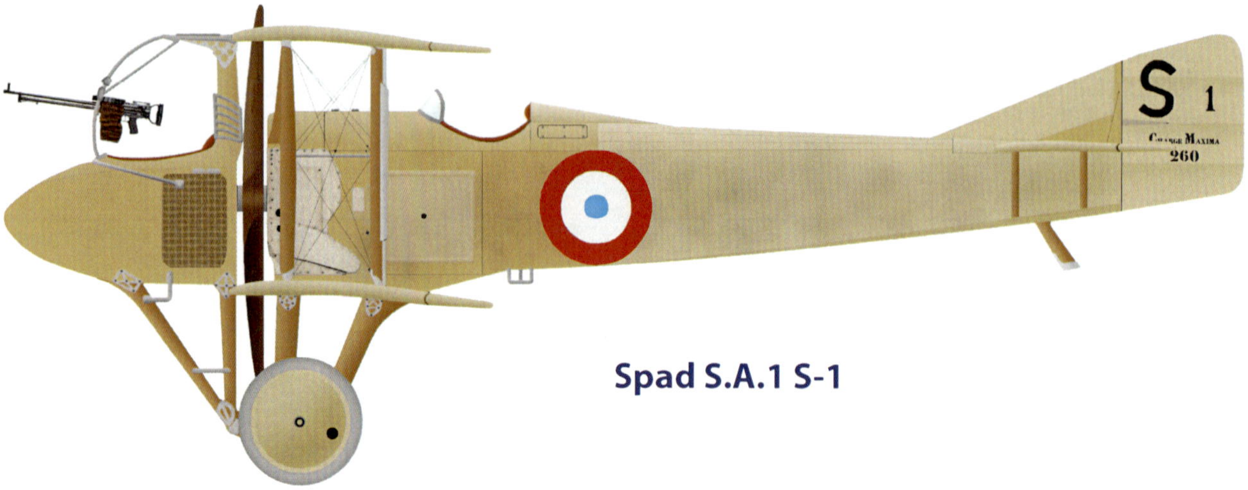

Spad S.A.1 S-1

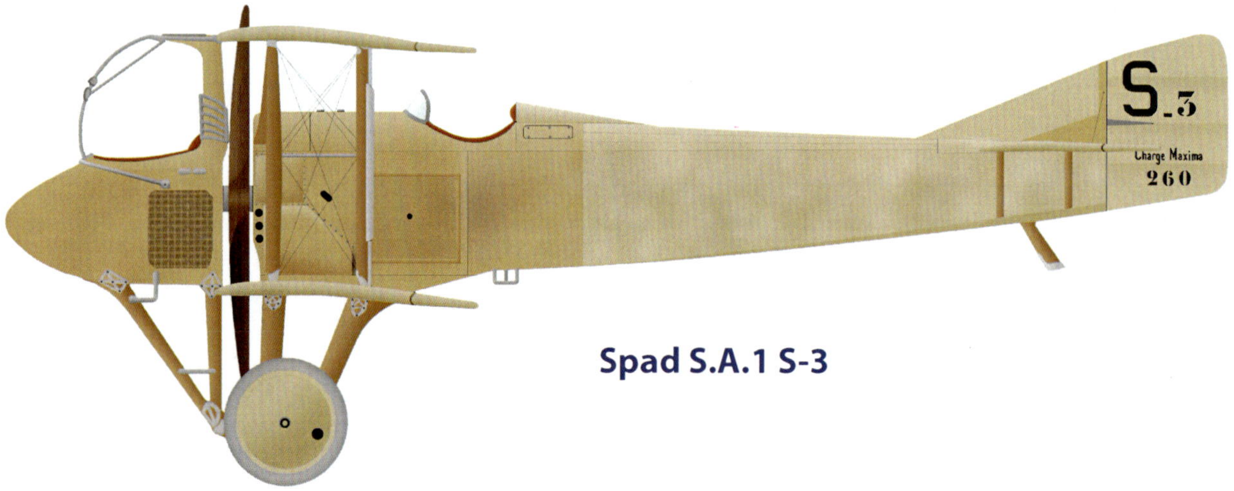

Spad S.A.1 S-3

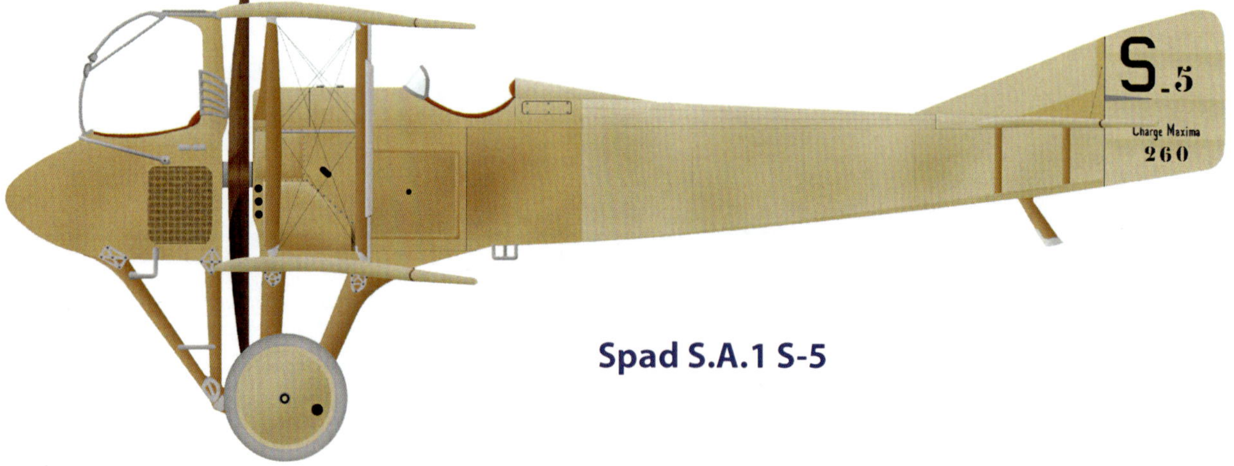

Spad S.A.1 S-5

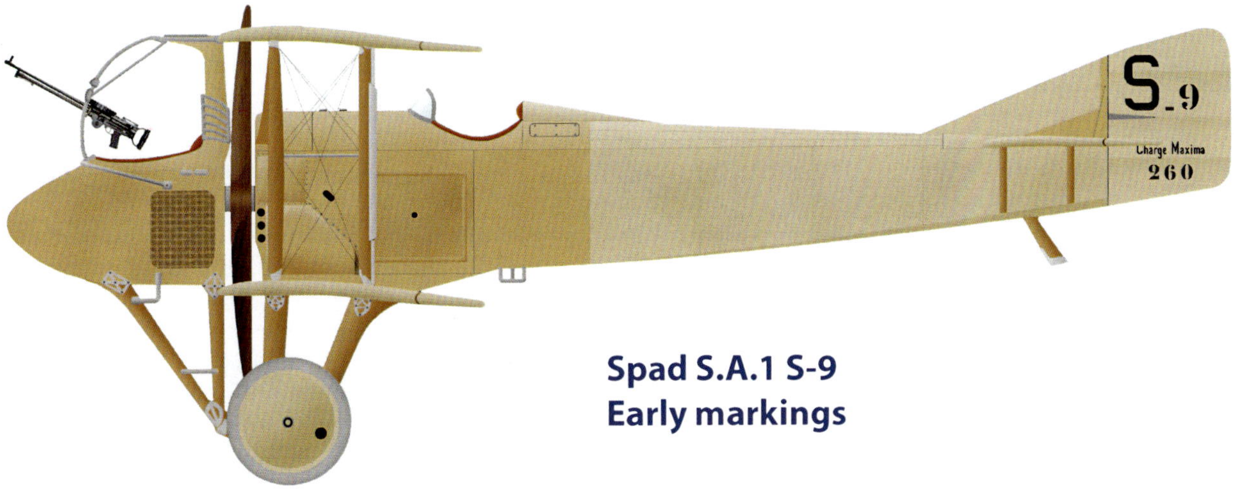

Spad S.A.1 S-9
Early markings

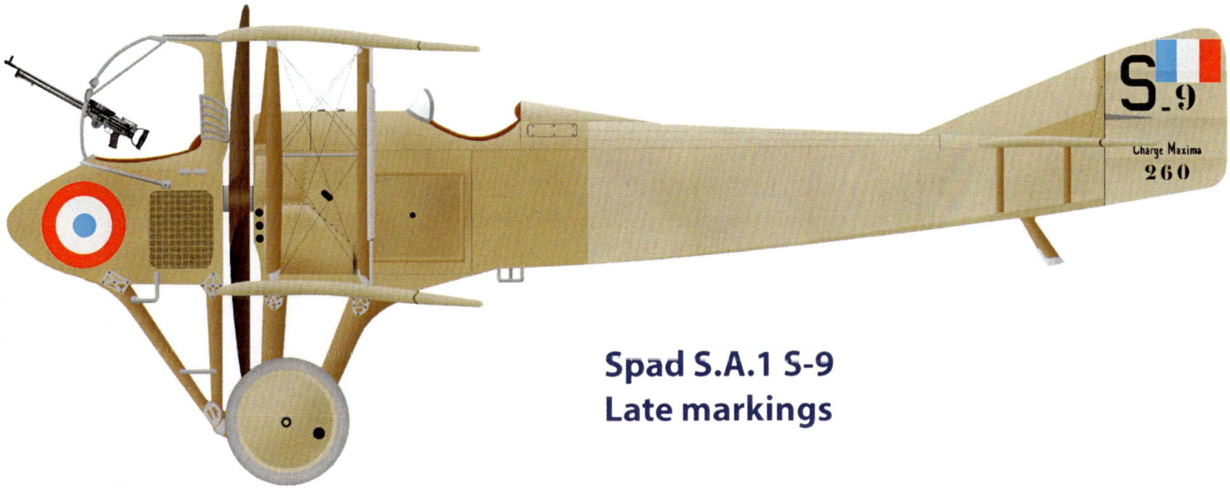

Spad S.A.1 S-9
Late markings

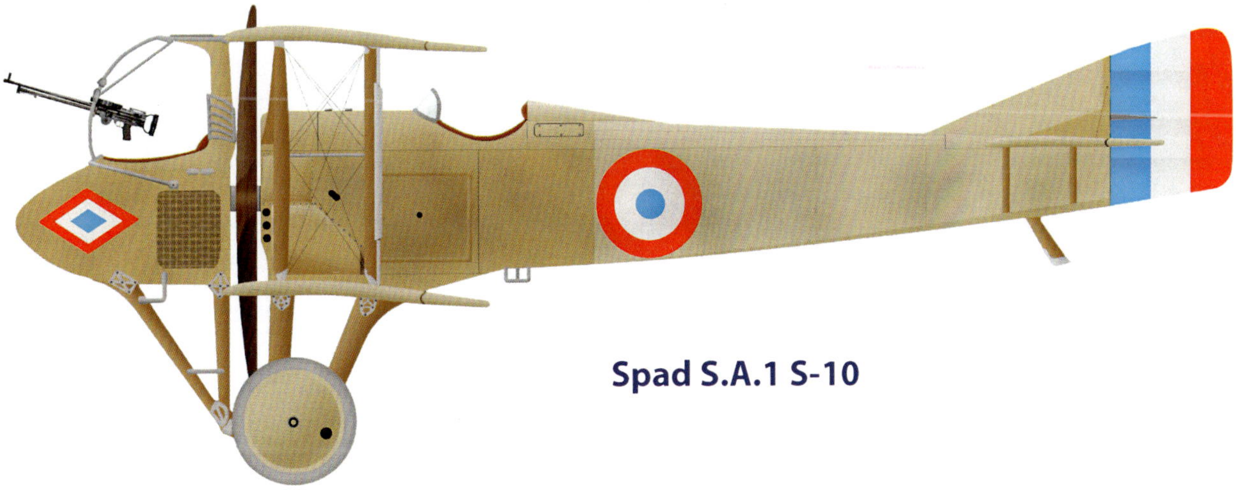

Spad S.A.1 S-10

Spad S.A.2 S-12

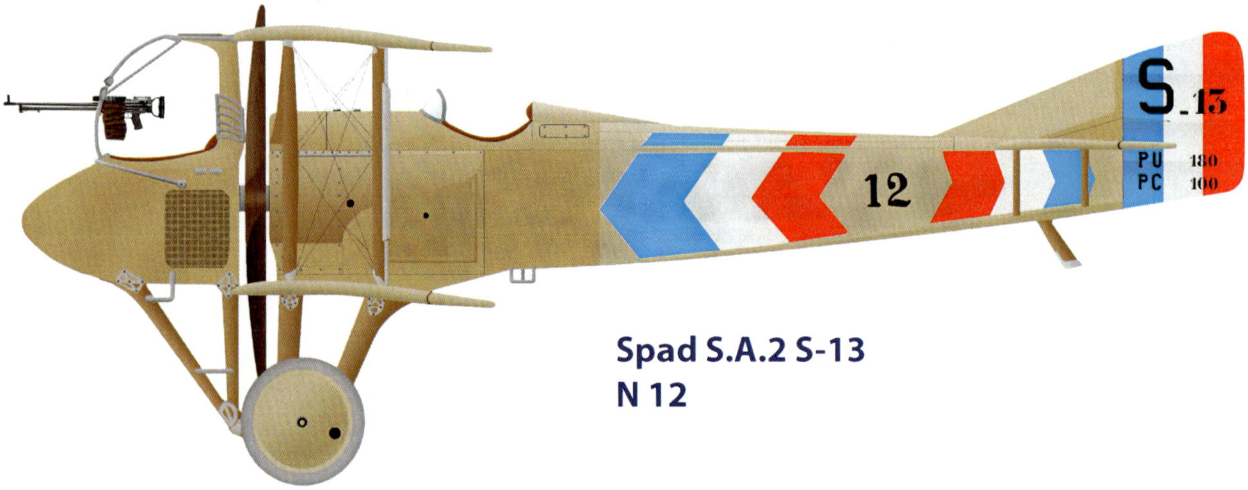

Spad S.A.2 S-13
N 12

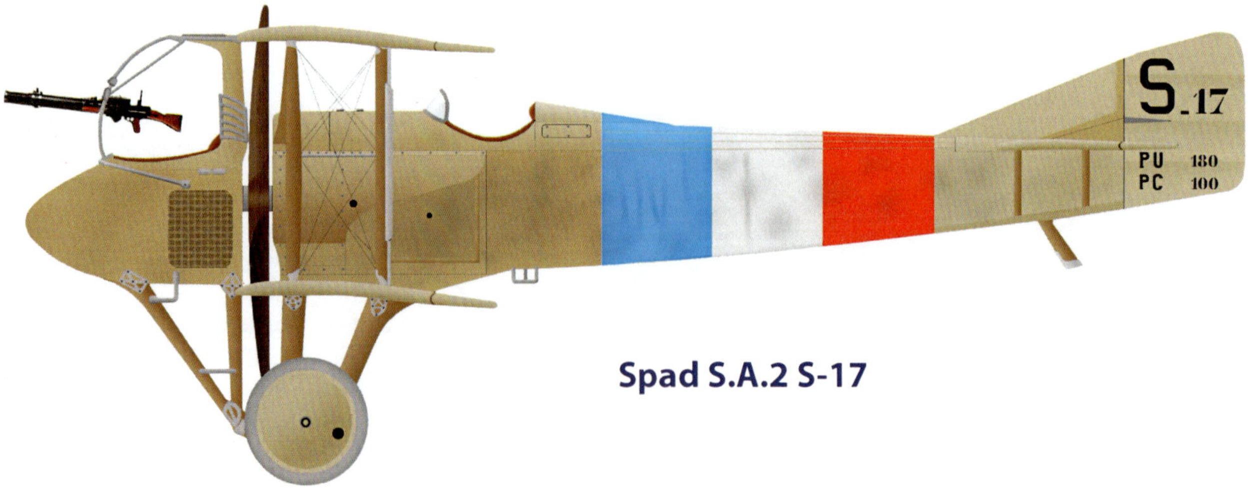

Spad S.A.2 S-17

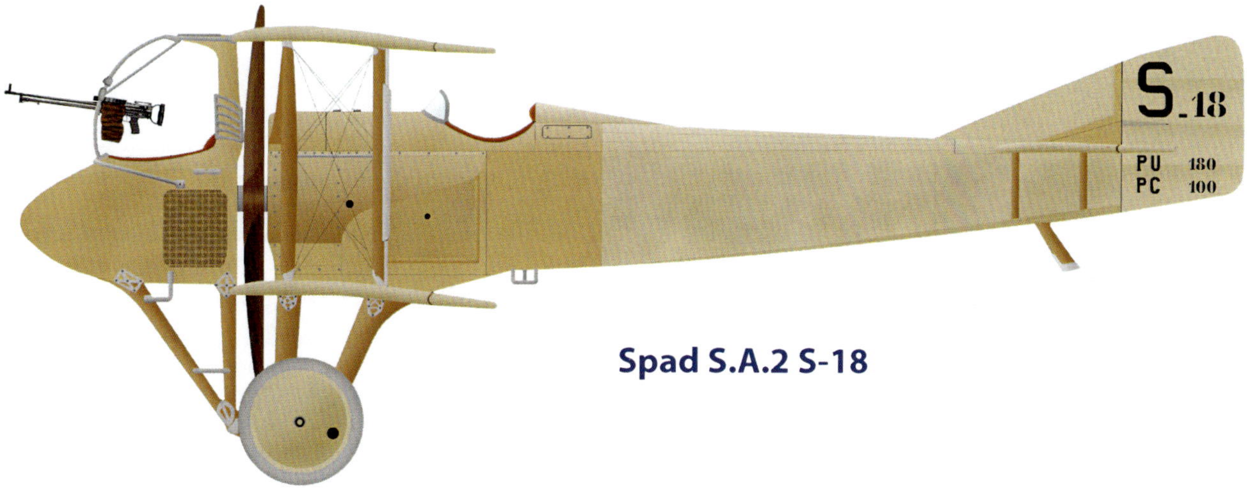

Spad S.A.2 S-18

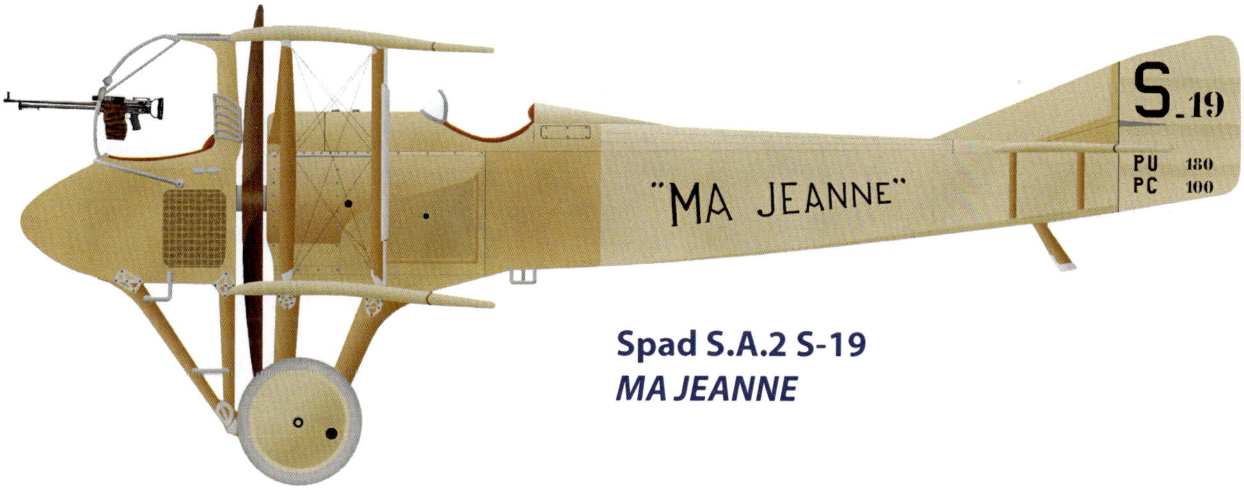

Spad S.A.2 S-19
MA JEANNE

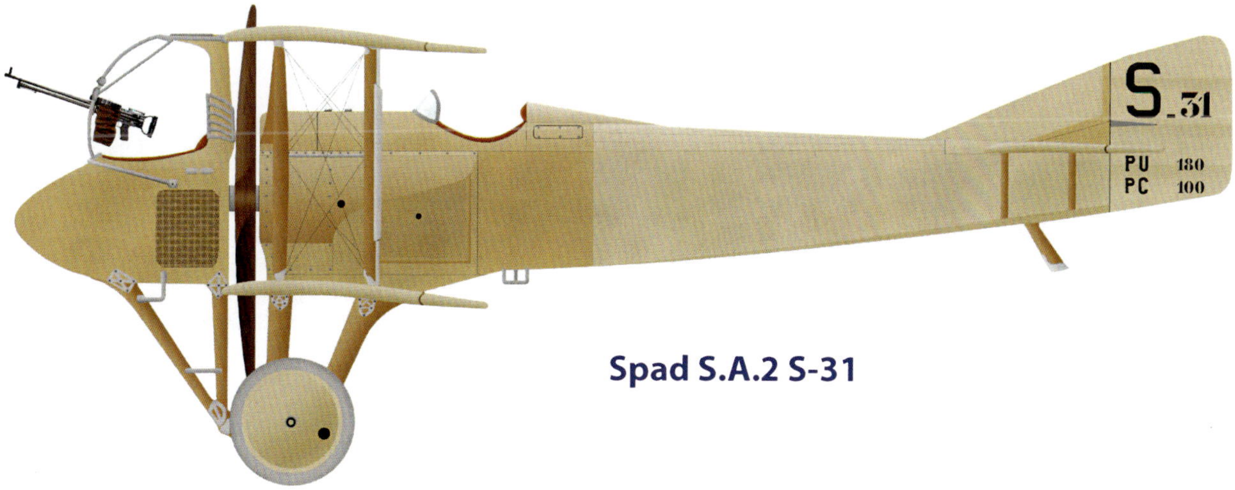

Spad S.A.2 S-31

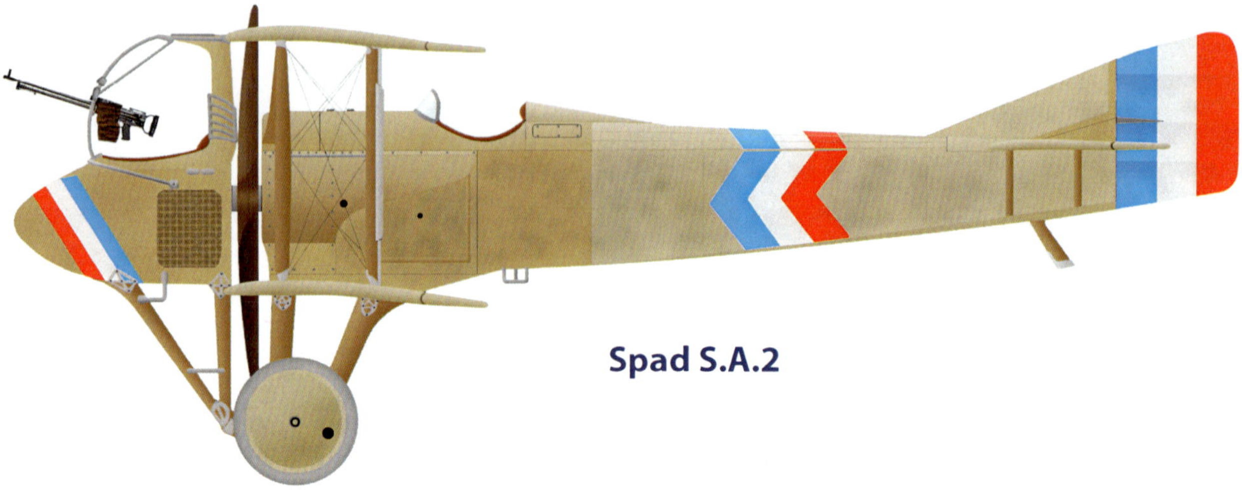

Spad S.A.2

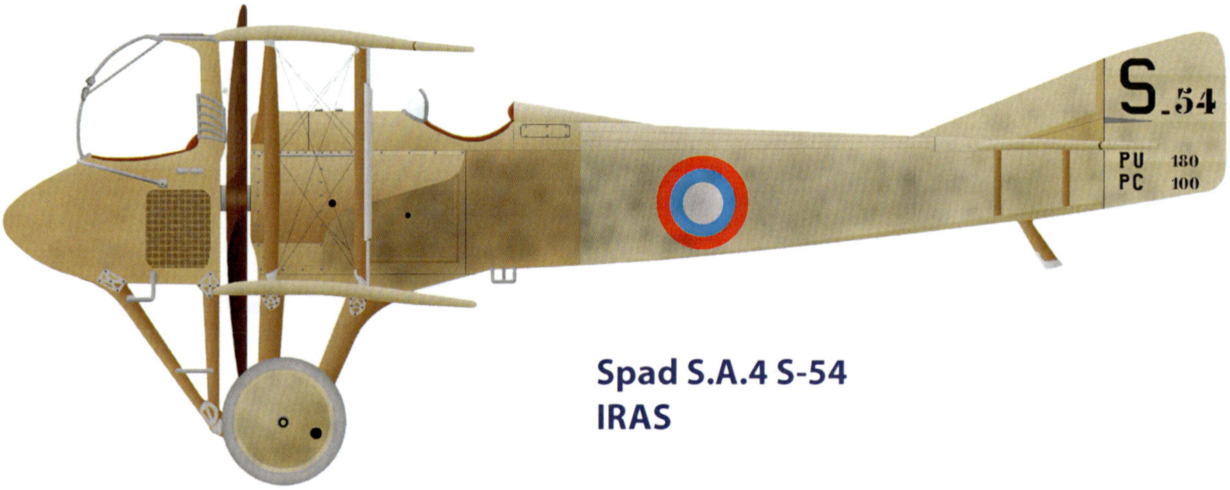

Spad S.A.4 S-54
IRAS

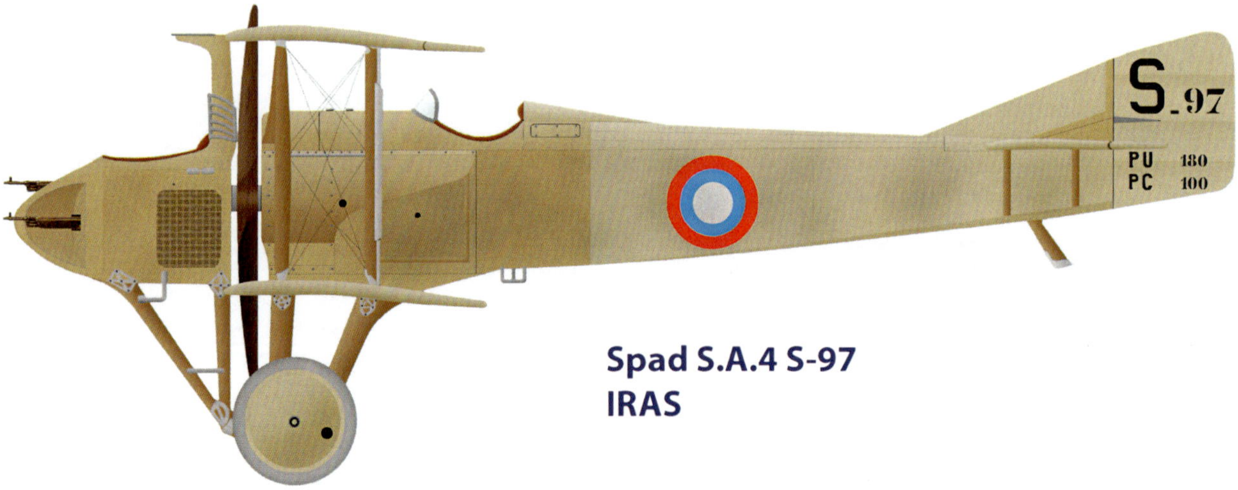

Spad S.A.4 S-97
IRAS

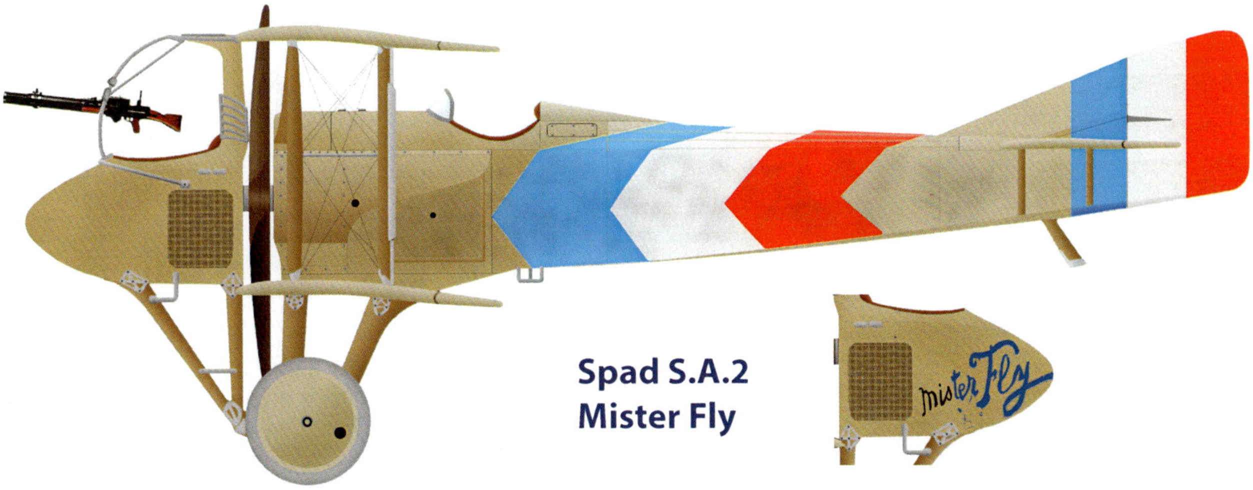

Spad S.A.2
Mister Fly

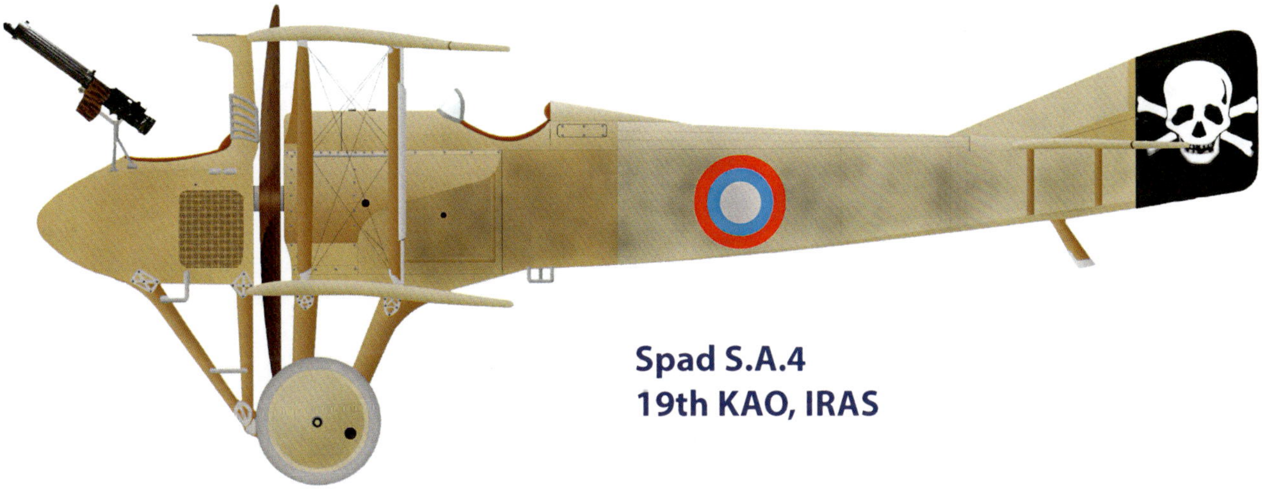

Spad S.A.4
19th KAO, IRAS

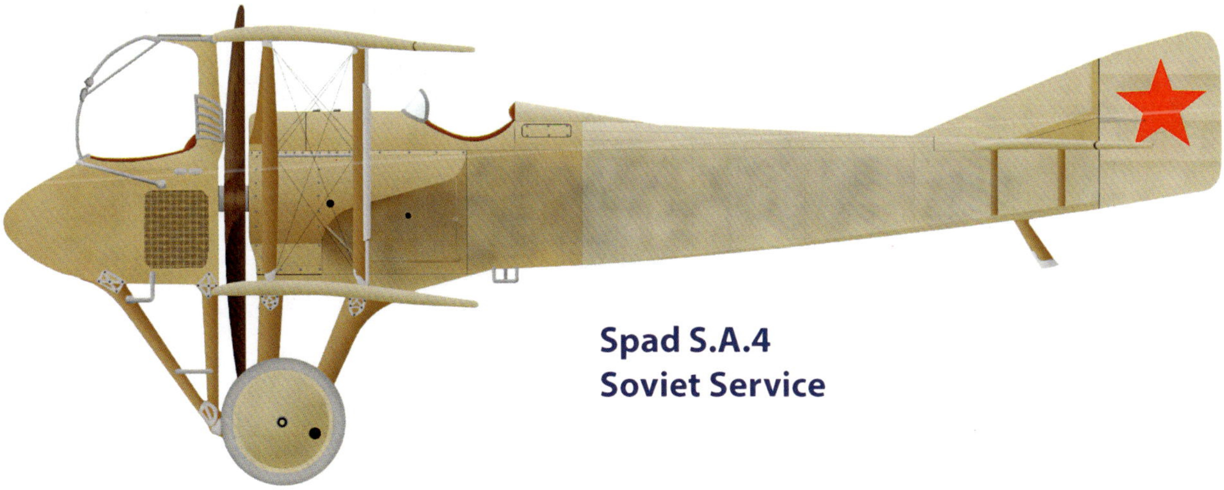

Spad S.A.4
Soviet Service

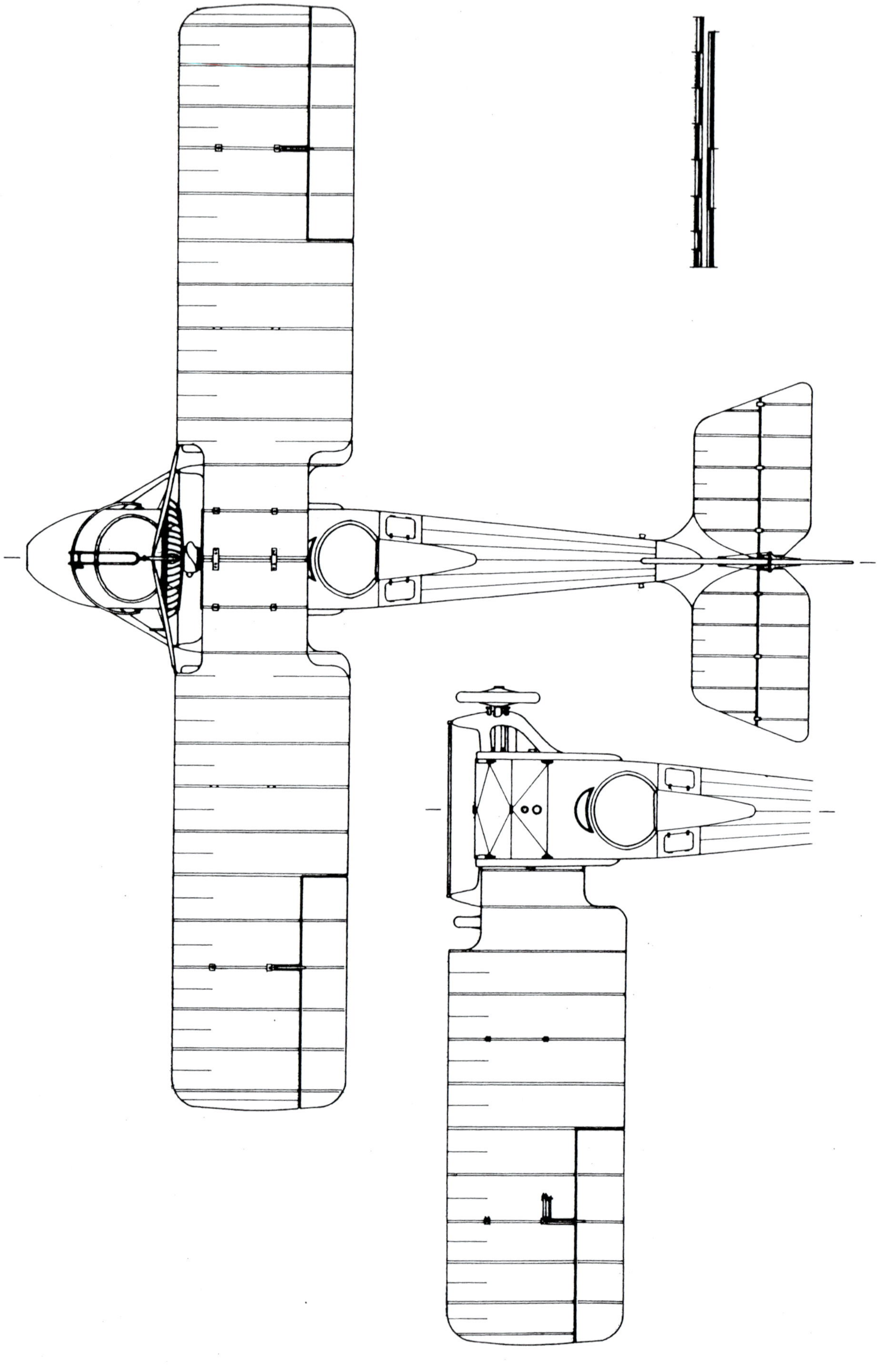

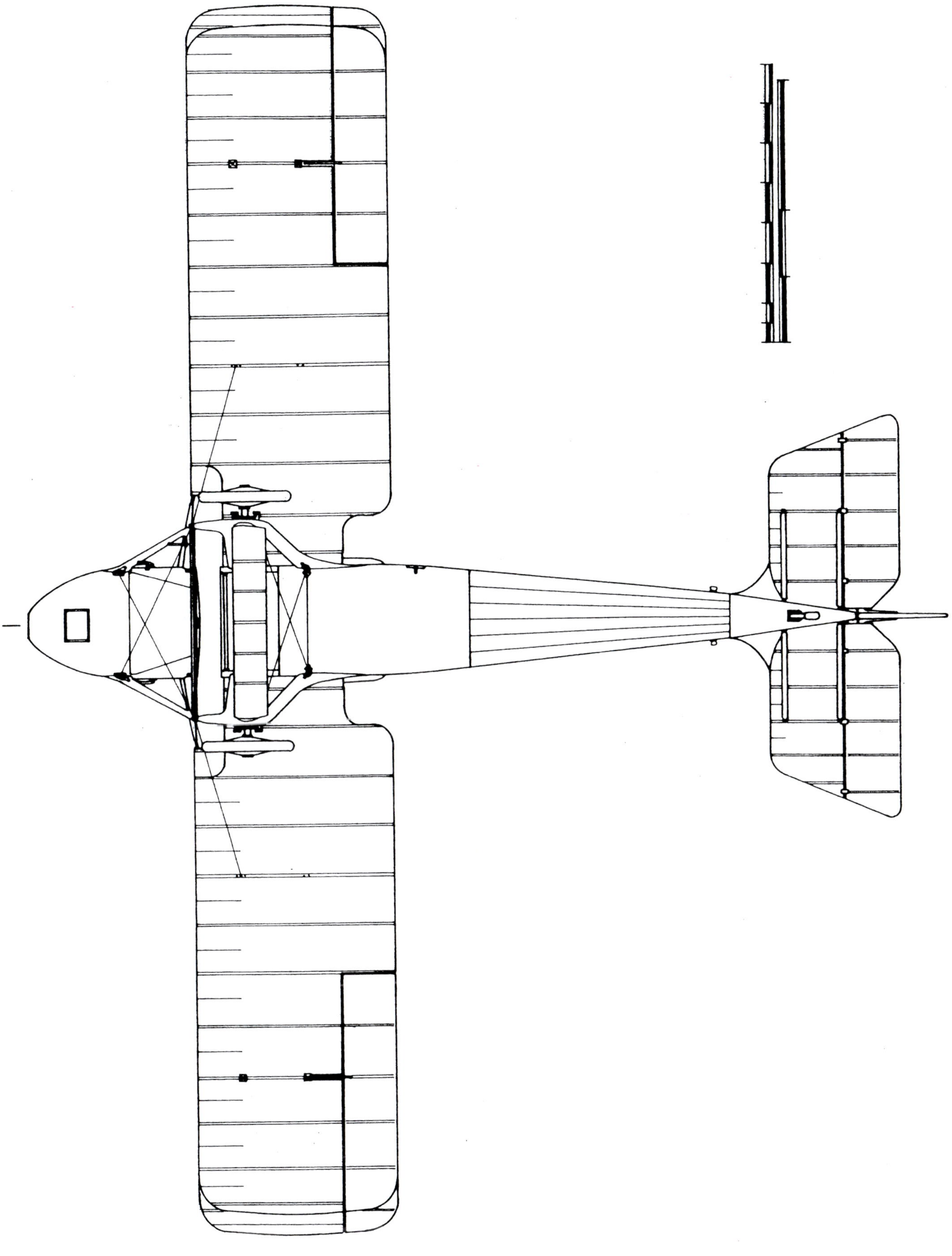

Spad S.A.2

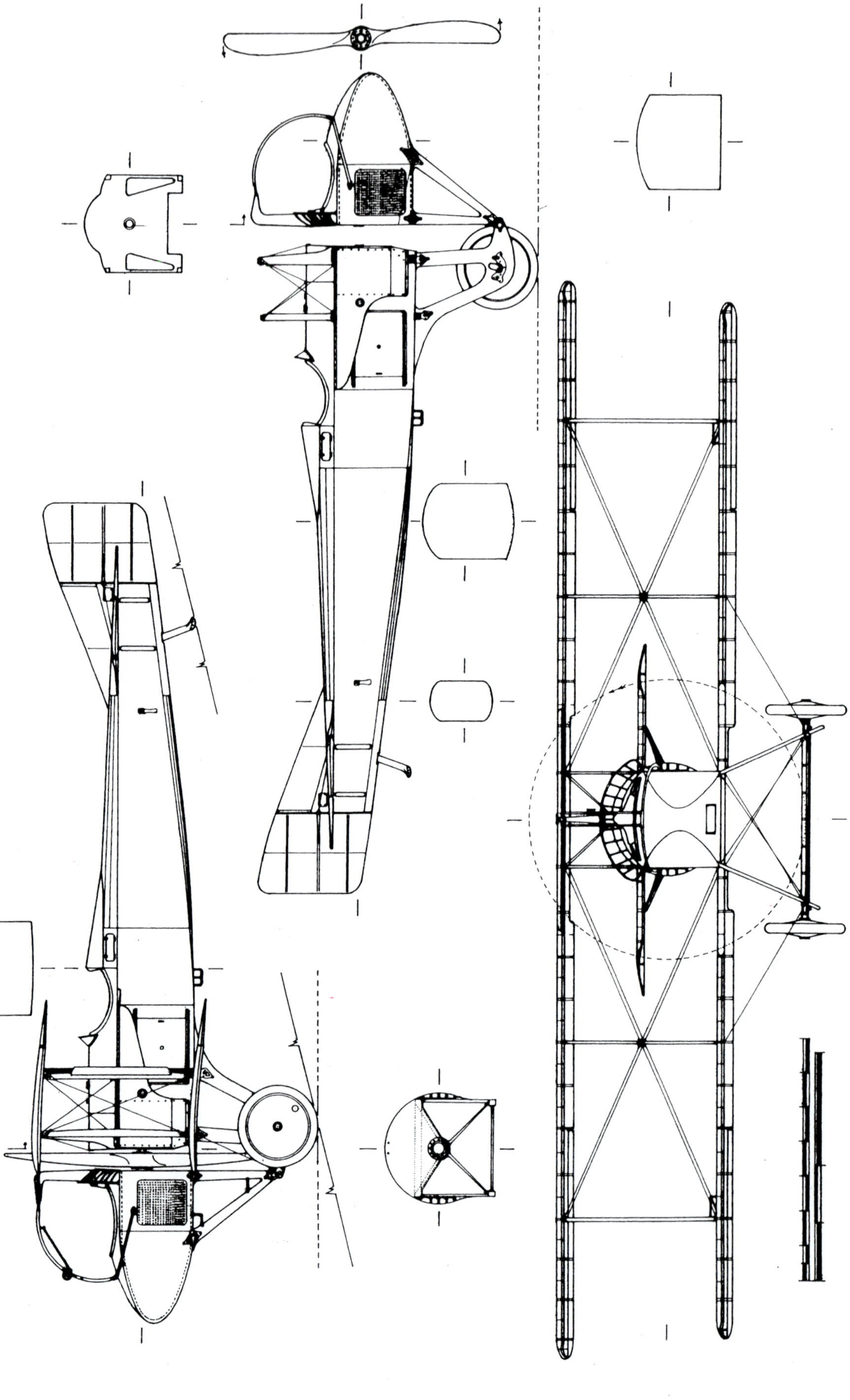

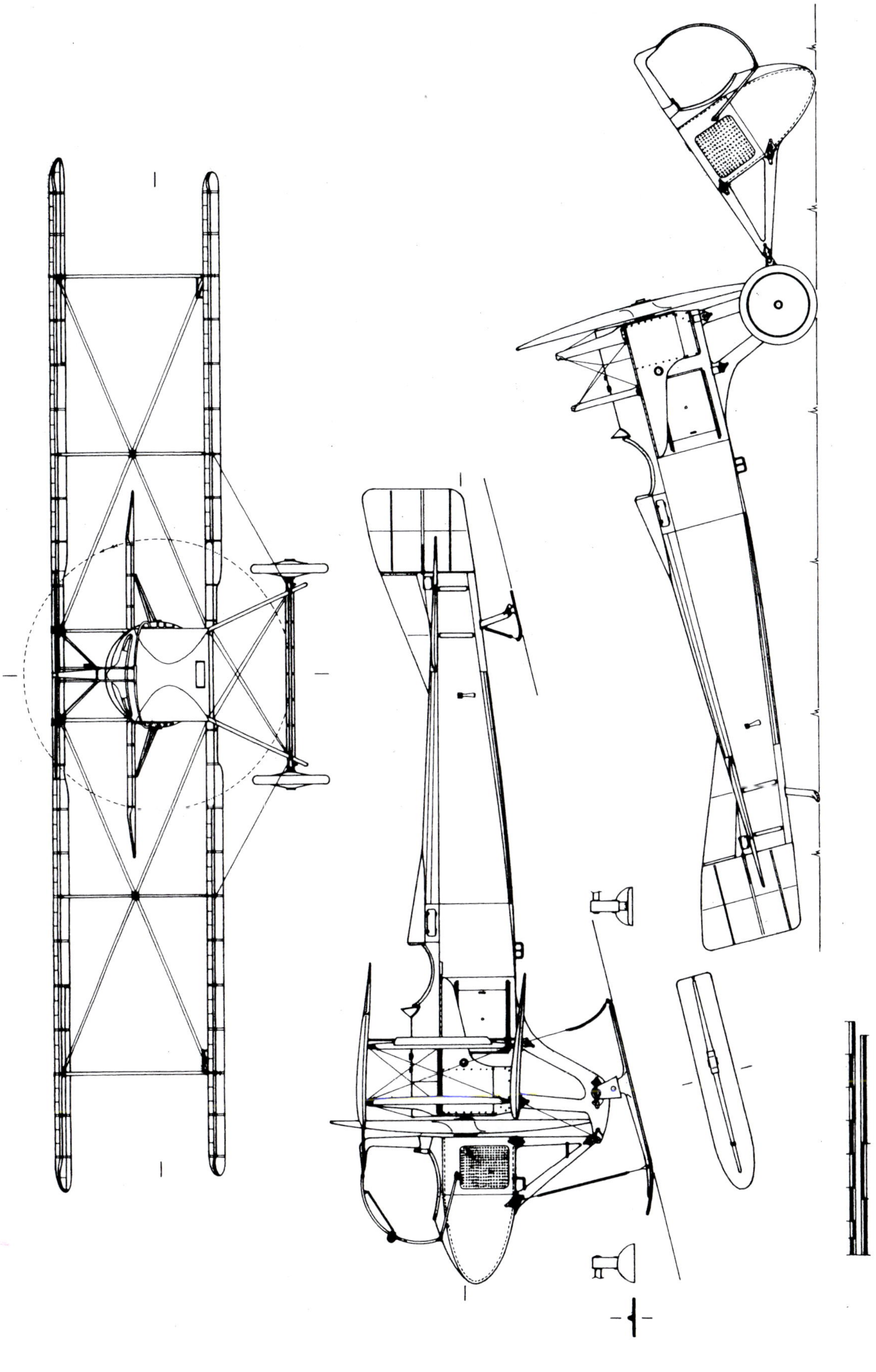

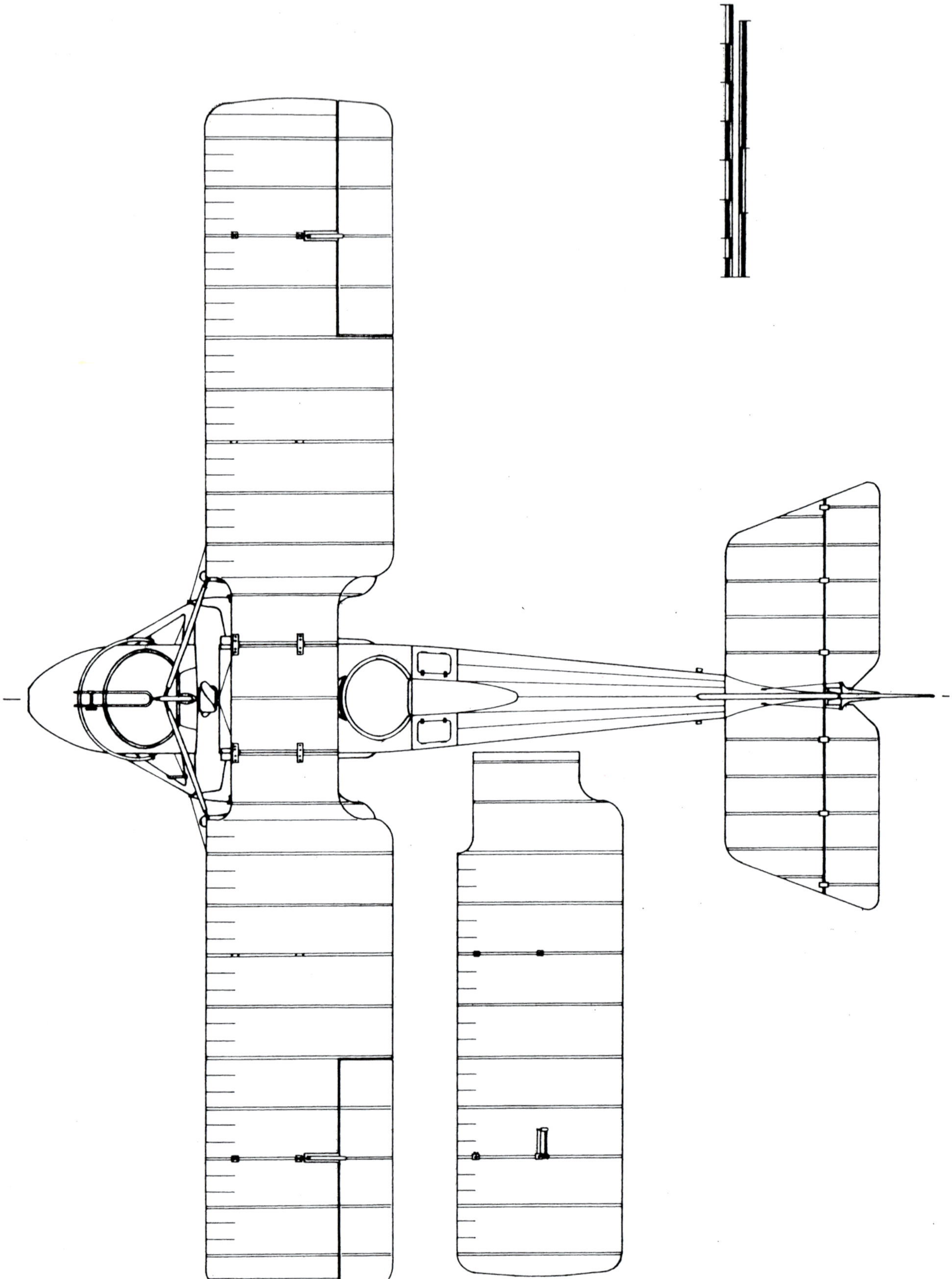

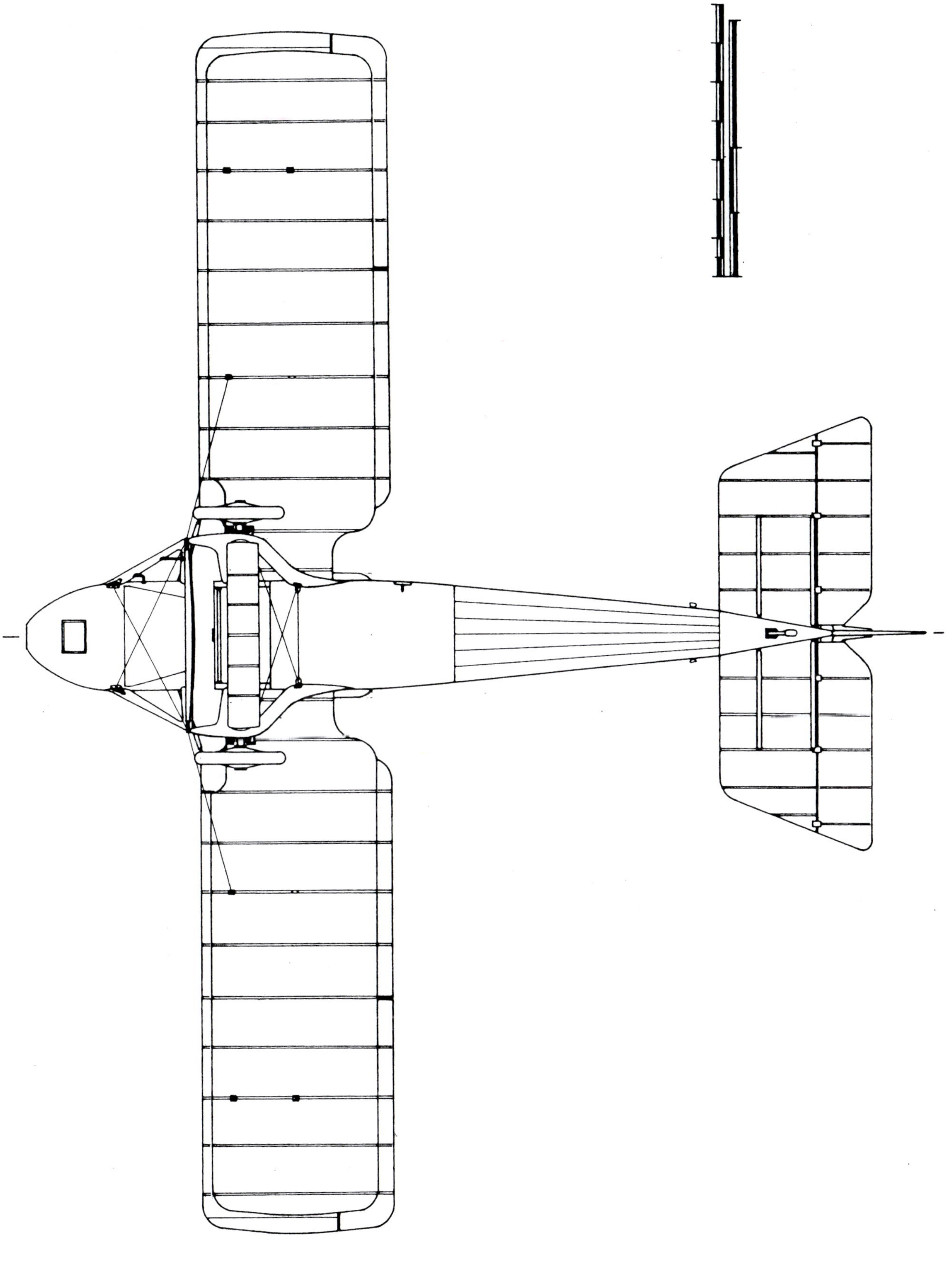

Left: S.A.3 was assigned to training use after evaluation.

Right: Pilots sitting in an S.A.2 cockpit.

Left: Damaged S.A.2 with nacelle down because there is no remaining support structure..

Made in United States
North Haven, CT
04 November 2026

21736873R00071